Introduction to Parallel Processing

INTERNATIONAL COMPUTER SCIENCE SERIES

Consulting editor   **A D McGettrick**   University of Strathclyde

SELECTED TITLES IN THE SERIES

Parallel Programming   *R H Perrott*

Concurrent Programming   *N Gehani and A D McGettrick*

Distributed Systems: Concepts and Design   *G Coulouris and J Dollimore*

Program Derivation: The Development of Programs from Specifications
   R G Dromey

Object-Oriented Programming with Simula   *B Kirkerud*

Parallel Processing: Principles and Practice   *E V Krishnamurthy*

Real Time Systems and Their Programming Languages   *A Burns and A Wellings*

Fortran 77 Programming   *T M R Ellis*

Prolog Programming for Artificial Intelligence (2nd Edn)   *I Bratko*

Introduction to Expert Systems (2nd Edn)   *P Jackson*

Computer Architecture   *M De Blasi*

The Programming Process: an Introduction using VDM and Pascal   *J T Latham,
   V J Bush and D Cottam*

Analysis of Algorithms and Data Structures   *L Banachowski, A Kreczmar and
   W Rytter*

Handbook of Algorithms and Data Structures in Pascal and C (2nd Edn)
   G Gonnet and R Baeza-Yates

Algorithms and Data Structures   *J H Kingston*

Principles of Expert Systems   *P Lucas and L van der Gaag*

Discrete Mathematics for Computer Scientists   *J K Truss*

Programming in Ada plus Language Reference Manual (3rd Edn)   *J G P Barnes*

Software Engineering (4th Edn)   *I Sommerville*

Distributed Database Systems   *D Bell and J Grimson*

Software Development with Z   *J B Wordsworth*

Program Verification   *N Francez*

Concurrent Systems: an Integrated Approach to Operating Systems, Database and
   Distributed Systems   *J Bacon*

Introduction to Parallel Processing

Bruno Codenotti
National Research Council, Pisa

Mauro Leoncini
University of Pisa

Translated by
Michele Re and Giovanni Cantatore

ADDISON-WESLEY
PUBLISHING
COMPANY

Wokingham, England · Reading, Massachusetts · Menlo Park, California · New York
Don Mills, Ontario · Amsterdam · Bonn · Sydney · Singapore
Tokyo · Madrid · San Juan · Milan · Paris · Mexico City · Seoul · Taipei

© 1993 Addison-Wesley Publishers Ltd.
© 1993 Addison-Wesley Publishing Company Inc.

All rights reserved. No part of this publication may be reproduced, stored in a retrieval system, or transmitted in any form or by any means, electronic, mechanical, photocopying, recording or otherwise, without prior written permission of the publisher.

The programs in this book have been included for their instructional value. They have been tested with care but are not guaranteed for any particular purpose. The publisher does not offer any warranties or representations, nor does it accept any liabilities with respect to the programs.

Many of the designations used by manufacturers and sellers to distinguish their products are claimed as trademarks. Addison-Wesley has made every attempt to supply trademark information about manufacturers and their products mentioned in this book. A list of the trademark designations and their owners appears below.

Cover design by Chris Eley and
printed by the Riverside Printing Co. (Reading) Ltd.
Printed in Great Britain at the University Press, Cambridge

First printed 1992

British Library Cataloguing in Publication Data
A catalogue record for this book is available from the British Library.

Library of Congress Cataloging in Publication Data
Codenotti, Bruno.
    Introduction to parallel processing / Bruno Codenotti, Mauro Leoncini.
       p. cm.
    Includes bibliographical references and index.
    ISBN 0-201-56887-X
    1. Parallel processing (Electronic computers) 2. Integrated circuits--Very large scale integration. I. Leoncini, Mauro. II. Title.
QA76.58.C62 1993
004'.35--dc20

                                                                                      92-29633
                                                                                          CIP

Trademark Notice
CRAY X-MP™ and CRAY-1™ are trademarks of Cray Research Inc.
CDC Cybor 205™ is a trademark of Control Data Corporation.
IBM 3090-VF™ is a trademark of International Business Machines Corporation.

Preface

Parallel computation is one of the richest and most fascinating fields of modern computer science. The book approaches the subject in a systematic way, giving precedence to the fundamental aspects without neglecting the technical and applicative points of view.

The book includes several examples stressing above all the algorithmic side. There is a complete and up-to-date bibliography, with respect both to textbooks and to papers published in the main scientific journals. Chapters 2-6 are supplied with bibliographic notes to help the reader venture into the vast referenced bibliography.

This textbook is mainly devoted to students in computer science, computer engineering, and mathematics. It is also directed at researchers in the field of parallel computation, and, in general, to all scholars seeking a deeper understanding of the theoretical aspects of parallel computation.

Readers' guide

The book is organized into six chapters. Chapter 1 is a general introduction. Chapter 2 is an overview of computation environments, models and sequential complexity. Chapter 3 is dedicated to models and complexity of parallel computation, while Chapters 4 and 5 contain the description of algorithms and of parallel architectures, respectively. Chapter 6 is an introduction to VLSI computation environments.

This book has been written as a whole and the reader will gain greater benefit by reading the chapters in their natural order, with possible omission of the introductory aspects. However, for those readers who are interested in only some of the topics, the following order is suggested: Chapters 1→ 3 → 4, 3 → 6. Chapter 6 refers to concepts and results discussed in Chapter 3; Chapter 4 contains the analysis of the algorithm in the models discussed in Chapter 3; Chapter 5 is independent of the rest.

Figures, examples, theorems, propositions and algorithms are numbered with reference to the chapter, and not to the section. Figures have their own numbering, while examples, theorems, propositions and algorithms follow a common one. Formulas are numbered separately.

Acknowledgments

We discuss in this book the fundamentals of parallel computation which is our past and present field of research. For this reason we would like to thank first of all Milvio Capovani who, besides contributing in many ways to our scientific education, has always guided and encouraged us. We also thank Mario Arioli, Roberto Bevilacqua, Dario Bini, Paola Favati, Giotto Fiorio, Grazia Lotti, Ornella Menchi, Claudio Montani, Chiara Puglisi, Giovanni Resta, Francesco Romani, Fabio Tardella, Andrea Tomasi and Paolo Zellini, who are part of a group which has been working for some time, in Pisa, in the field of parallel computation. The collaboration with them has been pleasant, stimulating and constructive. Special thanks to Giovanni Resta and Daniela Santangelo for their suggestions and for reading a preliminary version of this text. We must also thank the Addison-Wesley representative, Santiago Ojea, for his trust and for being very helpful while collaborating with us. Finally we would like to thank our wives: this book is dedicated to them.

Bruno Codenotti
Mauro Leoncini

Pisa, June 1992

Contents

Preface		v
Chapter 1	**Introduction**	**1**
1.1	Parallel computing	1
1.2	Technology	2
1.3	Classification of parallel machines	2
1.4	Interconnection structures	3
1.5	Parallel computing models and complexity measures	3
1.6	Other aspects of parallel computing	4
1.7	Structure of the book	5
Chapter 2	**Preliminaries**	**6**
2.1	Structure of a conventional computer	6
2.2	Basic computability theory	8
2.3	Functions, problems, languages	27
2.4	Concrete computational complexity	33
2.5	Complexity classes, reductions, completeness	47
	Bibliographic notes	56
	Problems	57
Chapter 3	**Parallel Computation Models**	**59**
3.1	Thesis of parallel computation	59
3.2	Parallel computations and algorithmic models	60
3.3	Boolean circuits	68
3.4	Parallel Random Access Machine (PRAM)	86

viii   Contents

	3.5	Structured models	106
	3.6	Distributed systems models	116
		Bibliographic notes	121
		Problems	122

Chapter 4 Parallel Algorithms 124

4.1	Basic techniques	124
4.2	Arithmetic algorithms	130
4.3	Combinatorial algorithms	137
4.4	Numerical algorithms	149
	Bibliographic notes	164
	Problems	164

Chapter 5 Parallel Architectures 166

5.1	Introduction to parallel architectures	167
5.2	Interconnection structures	187
5.3	Shared memory architectures	199
5.4	Message-passing architectures	204
	Bibliographic notes	205
	Problems	206

Chapter 6 VLSI Algorithms and Architectures 207

6.1	Introduction to VLSI technology	208
6.2	VLSI computing models	213
6.3	Area–time complexity	216
6.4	VLSI complexity of some computational problems	231
	Bibliographic notes	249
	Problems	250

Appendix A	Orders of magnitude	251
Appendix B	Number representation	252
Appendix C	Pidgin ALGOL language	254
Bibliography		257
Index		266

Notation

- $\log x$ is the base 2 logarithm of a positive number x;
- $\emptyset$ is the empty set;
- N is the set of natural numbers;
- Z is the set of integer numbers;
- R is the set of real numbers;
- $|x|$ is the absolute value of a scalar x;
- $|S|$ is the cardinality of a set S;
- $|f|$ is the cardinality of the codomain of a discrete function f;
- a div b is the quotient of the division of an integer a by an integer b;
- a mod b is the remainder of the division of an integer a by an integer b;
- $\|\cdot\|$ is a matrix or vector norm;
- $A = (a_{ij})$ defines a matrix A and the notation used for its elements;
- I is the identity matrix;
- $\rho(A)$, $\det(A)$, A^{-1}, A^T, A^m, $\mathrm{adj}(A)$ are the spectral radius, the determinant, the inverse, the transposed, the mth power, and the adjoint of a matrix A, respectively;
- $\mathrm{tr}(A)$ and $\mathrm{rank}(A)$ are the trace and the rank of a matrix A, respectively;
- $\mathrm{Cond}(A)$ is the conditioning number of a matrix A;
- $\lceil x \rceil$ is the lowest integer greater than or equal to x;
- $\lfloor x \rfloor$ is the greatest integer less than or equal to x;
- Σ^* is the closure of an alphabet Σ;
- $\deg(f)$ is the degree of a rational function f;
- the notation for orders of magnitude is given in Appendix A.

Chapter 1
Introduction

This chapter informally addresses the concepts that will be treated in the following chapters. The starting point is a preliminary investigation of the basic problems which are most frequently faced. This should help the reader to frame the setting and the structure of the book, beside stating its purpose.

1.1 Parallel computing

The term parallelism denotes the possibility of executing several operations (instructions) simultaneously. Nowadays there is a considerable amount of results on algorithms and parallel architectures, i.e. algorithms and architectures which make the parallel execution possible.

An intrinsic parallelism is present in many computational problems. While it is sometimes clear that the solution of many problems can be easily obtained by simultaneously executing some stages of the solving process, often a simple conversion of sequential algorithms into parallel algorithms does not provide the maximum attainable parallelism for a given problem. The fact that it is not always possible to transform the *best* sequential algorithms into the *best* parallel algorithms, i.e. there is no trivial correspondence between sequential and parallel computing, and the existence of parallel computing environments, open new horizons for scientific investigation.

The first parallel algorithms were developed at the beginning of the sixties, even though parallel architectures did not exist at that time. In that period, and in the following years, many researchers were fascinated by the challenge of solving problems assuming the existence of a *parallel environment*, without wondering about the applicability of their studies. These researches, which later became important in practice, led to the first investigations into the characteristics and properties of parallel computing.

1.2 Technology

Technology has been growing very rapidly during recent years. Different types of parallel computers have been designed, starting from the vectorial computer, up to the multiprocessor and the distributed system. These new architectures belong to the class (actually very large) of parallel computers. The design of parallel machines leads to a strong practical interest towards the comprehension of all aspects of parallelism, with a particular regard to the parallel solution of the main computational problems.

The advent of Very Large Scale Integration (VLSI) techniques has further augmented the interest towards parallelism from a different viewpoint: algorithms are designed in view of their hardware implementation. A very interesting feature of VLSI is its measure of "architectural complexity", i.e. the circuit area. Another attractive characteristic of VLSI is the possibility of having on a single circuit, at reasonable costs, a considerable number of computing elements cooperating for the execution of a given process.

Finally, VLSI is also innovative from a theoretical viewpoint: we will see that in order to obtain an appropriate computing model from it, it is necessary to take into account the "geometric" structure of computations, in addition to their logical structure.

1.3 Classification of parallel machines

The terminology used in parallel computing lacks uniformity even when dealing with the basic issues. This is because a great number of ideas, concepts and methods differing both from a logical and architectural point of view refer to the notion of parallelism. The characteristic which is common to all aspects of parallelism is the concept of *concurrence*, i.e. the simultaneous participation of entities in the quest towards a common goal. Developing the notion of concurrency implies abandoning both the traditional machine model (Von Neumann computer) and the sequential nature of its algorithms. The problem then arises of characterizing the new computing methodologies.

Already in 1966, some basic criteria had been identified to classify parallel computing from a programmer's point of view. Programming with a single instruction flow was called SIMD (Single Instruction Multiple Data), whereas programming with several instruction flows was called MIMD (Multiple Instruction Multiple Data). The SIMD and MIMD classes proved to be an important means of classification, since almost all parallel machines are actually included in them.

SIMD programming is also called *synchronous parallel programming*. There is a single program and all processors are constrained to execute the same instruction. The parallel flows which are automatically generated by an instruction on different data converge at the end of the instruction execution.

In the case of MIMD parallelism, the programmer has the possibility of controlling many instruction (process) flows and many data flows. All processes can be made of instructions belonging to different programs. This high flexibility makes MIMD parallelism more general than SIMD parallelism: a problem which is not characterized by a regular structure, but has a potential parallel exploitation, is appropriate for MIMD and not for SIMD. The greater flexibility must be paid for by the presence of problems which do not arise in the case of SIMD, in particular problems of data and process synchronization, and problems of allocation of processors to processes.

1.4 Interconnection structures

One of the main obstacles to the design of parallel computers resides in the difficulty of efficiently implementing hardware and/or software communication mechanisms among the different computing elements. If all the elements are connected by a shared resource, this becomes the *bottleneck* of the system with a consequent slower execution and worse performance. Hence, the interconnection problem should be addressed with the adoption of *dedicated* links. In this framework, a parallel architecture can be seen as a set of processors connected by links for exchanging messages. In order to analyze the properties of interconnection structures, it is convenient to adopt the notion of a graph. In fact, it is possible to establish a correspondence between the nodes of a graph and computing elements, as well as between arcs and links. The use of this correspondence makes the study of the topological properties of the structure much easier. It follows that if the focus is on the geometry of the connections, a graph is a natural representation of the architecture.

1.5 Parallel computing models and complexity measures

Computational complexity is a main field of computer science devoted to the determination of the minimal quantity of resources sufficient to solve a computational problem. The investigation is usually carried out in two different ways: one attempts to evaluate the quantity of resources necessary to solve a problem (complexity lower bound) and the other one analyzes the way algorithms use their resources (complexity upper bound). If one wants to evaluate algorithm performance and, when applicable, to compare it to the best possible performance, it is necessary to introduce computing models highlighting the fundamental resources and their relationships. It is then possible to define an optimum algorithm as the one making the *minimum* use of resources.

The models of parallel computation that we consider are simple enough to be mathematically treatable; however, their features are sufficiently close to that of actual parallel machines. In this way the analyses carried out on the

4 Introduction

models give results that can (to a greater extent) be applied to the machines.

An apparent contradiction regarding the complexity of parallel computations is the fact that many computing models do not consider the costs due to data access and data exchange. Neglecting these costs is in certain cases an excessive simplification of reality. On the other hand, the lower complexity bounds obtained by neglecting these costs are of course still valid, when these costs are present. It follows that the study and the analysis, even if performed on simplified models, give useful indications, provided that a model expressing some of the peculiarities of parallel machines is used. For this reason, many complexity analyses are performed on models neglecting some costs.

One of the main objectives of computational complexity is to determine bounds to the cost required by computations, independently of technical details. Hence, it is necessary to define how to measure the performance of parallel algorithms independently of the specific architecture. Chapter 3 will introduce some of the main parallel computation models, (Boolean and arithmetic) circuits and parallel random access machines (PRAM), together with an "algorithmic" model, which is less formal than the preceding ones, but more useful to analyze the performance of algorithms. This parallel computing model has been widely used, especially for the early complexity analyses. The following rules describe the criteria to perform an analysis using such a model:

1. any number of processors can be used at any time;
2. each processor can execute one (arithmetic or logic) operation in a unit of time;
3. data access has no cost;
4. communication among processors has no cost.

One of the most important aspects of computation models is their generality. The notion of a computation graph of a parallel algorithm is the unifying element of our approach. This represents the main principle of the analysis that will be done in Sections 3.2, 3.3 and 3.5, and, for VLSI complexity, in Sections 6.2 and 6.3.

1.6 Other aspects of parallel computing

This book is particularly dedicated to the algorithmic and complexity aspects of parallel computing, especially on computation models with a common memory, and circuits. Chapter 5 introduces the basic concepts of parallel architectures. An introduction to the theory of distributed parallel architectures is found in Leighton [Le91]. For more concrete aspects a good starting reference is the collection [TV86]. A deeper understanding can be gained by reading specialized reviews and conference proceedings (e.g. *Journal of*

Parallel and Distributed Computing and *International Conference on Parallel Processing*).

Another important aspect of parallel computing which is not touched in this volume is the real programming of parallel machines. This involves a consideration of the problems of developing compilers for parallel languages and the study of a program's correctness (semantics of concurrency). Many researchers, especially in Europe, have studied these aspects of parallelism. As a starting reference, one can consult the proceedings of the annual PARLE (Parallel Architectures and Language Europe) conference, published as volumes of the *Lecture Notes in Computer Science* series (published by Springer–Verlag), and specialized reviews.

1.7 Structure of the book

The main purpose of this book is to introduce the fundamental concepts of parallel computing. At the same time, we will not neglect the concrete aspects. Particular attention is devoted to the fundamentals, which are also important for a clear understanding and an adequate classification of more technical/practical results.

The peculiarity of this book rests in attempting a coherent synthesis between theory and practice. Chapter 2, which introduces some general concepts of theoretical computer science, has been included to make the volume self–contained. At the same time, it puts the description of sequential models of computation before the treatment of parallel models of computation. Chapter 3 handles areas which are inherent to the complexity of parallel computation and which in most cases are still under investigation. This is done by means of the definition of parallel machine models and of a classification of problems based on their "difficulty".

Chapter 4 analyzes some classes of parallel algorithms, e.g. arithmetic, combinatorial and numerical algorithms, using the most appropriate model. Chapter 5 treats the real parallel architectures. The approach is to abstract towards the previously introduced models, thus highlighting their usefulness as a tool for the analysis and synthesis of algorithms. Finally, Chapter 6 is dedicated to the description of VLSI computing, with particular attention to the study of the area–time complexity, and to the direct implementation of algorithms by circuits.

Chapter 2
Preliminaries

The purpose of this chapter is both to provide the basic knowledge necessary to study the fundamentals of parallel computing and to understand its basis. It has been included to make the volume as self-contained as possible. Section 2.1 briefly shows the structure of a conventional computer. Section 2.2 lists the basic notion of computability and introduces the main sequential computing models, i.e. computers themselves, as the conventional ones which execute one instruction at a time. Section 2.3 formally introduces the problem concept. Section 2.4 deals with the fundamental ideas for the complexity analysis of problems. Section 2.5 summarizes this subject, while introducing the basic ideas for structuring a problem in classes of complexity.

2.1 Structure of a conventional computer

A conventional computer, the structure of which is shown in Figure 2.1, has three main elements: the *central processing unit* (*CPU*), the *memory* and the *input/output* (*I/O*) *devices* (or peripherals). They are the hardware of the computer, i.e. that part of a computer which is physically built.

The "heart" of a computer is the CPU–memory pair, which characterizes its behaviour. The memory, subdivided into independently addressable locations, contains the program which, in turn, specifies the sequence of instructions to be executed, as well as the data to be used. The CPU can execute instructions belonging to a limited set. Moreover, it has a given number of registers, used in order to execute instructions. Finally, the I/O devices are useful to communicate in both directions with the outside world, i.e. receiving and sending data. We will use the word *status* to denote a global description of memory, CPU registers and I/O devices.

A computation is performed in the following way. The CPU reads the program instructions, one at a time, and executes the related operations. Each single instruction causes a change in the status of the computer. Thus, a *computation* may be seen as a series of states $S^{(0)}$, $S^{(1)}$, $S^{(2)}$, ..., where $S^{(0)}$

Structure of a conventional computer

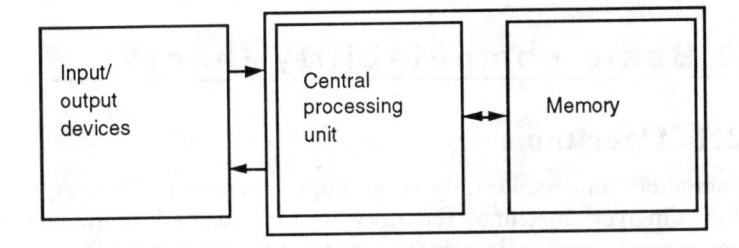

Figure 2.1 Structure of a conventional computer.

is the status of the computer at the beginning of the computation and $S^{(i+1)}$, $i = 0, 1, 2, ...$, is obtained by $S^{(i)}$, following the execution of a well–defined instruction belonging to $S^{(i)}$. Note that the status sequence can be either finite or infinite. In the former case we say that the computation *converges*, in the latter case we say that the computation *diverges*. Diverging computations correspond to the well–known situation of "looping programs".

Computers structured as a conventional computer are historically called *Von Neumann machines*. The main characteristics of the organization and behaviour of a conventional computer can be summarized as follows:

1. Programs and data reside together in memory.
2. The execution of a program implies an exchange of information (data and program instructions) between CPU and memory.
3. Program and data acquisition implies receiving information from the outside world.
4. Output of results implies sending information to the outside world.
5. Instructions are executed in sequence.

Three main kinds of event may occur in a computer:

(a) instruction execution;
(b) information storing;
(c) information exchange.

Event (a) occurs in the CPU, which represents the place where time is spent. Event (b) occurs in memory, which represents the device where storage is performed. Finally, event (c) occurs in the links between CPU and memory, and between the CPU–memory system and the peripherals.

We have identified three crucial resources: *storage* (in memory), *computing time* (in CPU) and *communication time* (through the links). It follows that a program for a conventional computer, as shown in Figure 2.1, can be considered a "good" one if it makes good use of memory and time resources.

2.2 Basic computability theory

2.2.1 Algorithms

A computer can execute only very simple operations. Typical examples, included in every machine's repertory, are addition and subtraction, as well as operations moving information between memory and registers, and between memory and I/O devices. The execution of complex activities is possible because of the great speed available for executing these simple operations. Complex activities must then be described in terms of elementary operations. An *algorithm* is such a description. The structure of a conventional computer and above all the sequential nature of its computations made of discrete steps corresponds to the idea of an algorithm. A precise definition of this concept was derived prior to the advent of modern computers, by defining primitive computational models which contain all the basic ideas of automated calculus. Section 2.2.2 will show the main sequential computing models. First, however, we must make some general observations about algorithms.

The concrete expression of an algorithm implies a communication *language* and further a *lexicon* and a *syntax*. Recall that, given an alphabet Σ,[1] Σ^* denotes the set consisting of all the *finite–length* (greater than or equal to zero) strings of symbols belonging to Σ. A language L on an alphabet Σ is any given subset of Σ^*. In practice, artificial languages with different degrees of formalism are used to render algorithms. A language conceived to render algorithms is called a *programming language*. Any algorithm codified with a programming language is called a *program* (more precisely the strings of the language are called programs). The nature of the formalism can be very different, as shown in the following two examples:

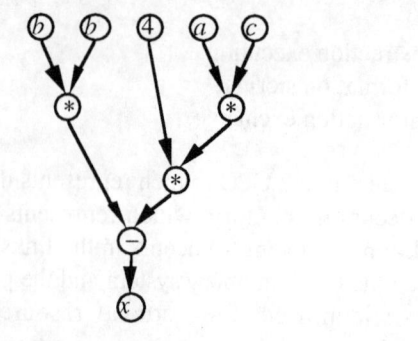

$$\text{(1)}$$

$$x \leftarrow b*b - 4*a*c \tag{2}$$

---

[1] An alphabet Σ is a set of symbols, and a string on Σ is a concatenated sequence of symbols belonging to the alphabet.

Two intuitive requirements apply to every programming language. First the alphabet Σ on which L is defined must be finite; second, for any given string $w \in \Sigma^*$, a rule exists to decide wether $w \in L$. Pidgin ALGOL, described in Appendix C, is an example of a language specified by a lexicon and a syntax with a sufficient level of formality. When an algorithm is not written to be executed by a machine, but, for instance, to transmit ideas to other persons (algorithms described in a book are often of this kind), the need for formality is not so strong. Still the need to avoid ambiguities must not be neglected.

Along with the linguistic problem there is the algorithm *semantics* problem, that is, the problem of its meaning. In this book, we will consider two different ways to give meaning to an algorithm: the *operational* and the *functional* way.

- Operational semantics describes the results of the execution of an algorithm, that is, the computations that it will generate. The nature of a computation depends on the machine for which the algorithm has been written (this is due to the fact that modifications are made on the state of the machine itself). A computation (see Section 2.1 on ordinary computers) consists of a sequence $S^{(0)}, S^{(1)}, S^{(2)}, ...$, of machine states (or *configurations*). $S^{(0)}$ is the initial state of the computation and $S^{(i+1)}$, $i = 0, 1, 2 ...$, is obtained from $S^{(i)}$ by executing an instruction, i.e. a *step* of the computation.
- Functional semantics gives meaning to an algorithm by considering its input/output behaviour. Note that a function f from a set D to a set C (one writes $f: D \rightarrow C$) is a prescription that associates *at most* one element $c \in C$ to every element $d \in D$. If there exists an element c corresponding to d, f is said to be defined at d (or that $f(d)$ is defined), and one writes $c = f(d)$. If a function is defined at every domain point, it is called *total*, otherwise it is called *partial*.

Operational semantics provides the means to introduce an important algorithm classification based on computations.

Definition 2.1
A computation $S^{(0)}, S^{(1)}, ..., S^{(i)}, S^{(i+1)}, ...$, is *deterministic* if it is completely determined by the initial state $S^{(0)}$, i.e. if, for each i, $S^{(i+1)}$ is a function of $S^{(i)}$.

Note that all program data (if any) are included in the description of $S^{(0)}$. The machine status in a conventional computer (this concept will become clearer when dealing with other computing models) also includes a description of I/O devices, where a read operation transfers data into main memory. It follows that, provided an algorithm produces deterministic computations, it is always possible to associate a function (either partial or total) with it.

Definition 2.2
Let $S_1, S_2, ..., S_k$, be k distinct states. A computation $S^{(0)}, S^{(1)}, ..., S^{(i)}, S^{(i+1)}, ...$, is called *probabilistic* if there exists at least a status the successor of which can only be determined with a given probability, i.e. there exists an index i such that $S^{(i+1)} = S_j$ with probability $p_j, j = 1, ..., k, \sum_{j=1}^{k} p_j = 1$.

Definition 2.3
Let $S_1, S_2, ..., S_k$, be k different states. A computation $S^{(0)}, S^{(1)}, ..., S^{(i)}, S^{(i+1)}, ...$, is called non–deterministic if there exists at least an index i such that $S^{(i+1)}$ is an element of $\{S_1, S_2, ..., S_k\}$.

A non–deterministic computation has no stochastic law determining the next state. Probabilistic computations and non–deterministic computations have similar definitions. However, they represent different aspects of computations.

In a probabilistic machine (i.e. a machine that can produce probabilistic computations) state transitions succeed on the basis of a certain law. In other words a probabilistic machine is a computer with an attached random number generator, generating numbers uniformly distributed on a sufficiently large set. As an example, a probabilistic computation in which each state has at most two successors (each one with the same probability) can be generated from a machine connected with a device storing the results obtained by repeatedly throwing a (perfectly balanced) coin. An algorithm referring to probabilistic devices is called a probabilistic algorithm.

On the other hand, the concept of non–determinism is useful to study the computational complexity of problems for which the best known solution strategy is the complete search, i.e. the generation of all possible solutions and their verification. A non–deterministic machine solves these problems generating in a single step the "right" configuration (if it exists) and verifying that it is actually a solution to the problem. A non–deterministic machine can be defined as a machine always making "lucky" choices. Naturally, such a device is only an ideal one. Let us consider the following example.

Example 2.4
Consider the problem of determining if a positive number x is not prime. The following "algorithm" could solve the problem by using a non–deterministic machine. Let y and z be two positive integers non–deterministically chosen such that $1 < y, z < x$; then the answer to the problem is yes if and only if x equals the product yz. In fact, if $x \neq yz$ we can be sure that two numbers y and z with the above properties do not exist because the machine is always making lucky choices.

The previous algorithm is non–deterministic because it uses a non–deterministic criterion for the choice.

Another possible way of considering non–determinism is to imagine a non–deterministic machine as a machine following in parallel all the possible alternatives. In the case of the previous example, there is only a non–deterministic step, i.e. the choice of y and z. The number of possible alternatives is $(x - 1)(x - 2)/2$, which represents the number of possible (unordered) pairs of integer numbers y and z satisfying the constraint $1 < y$, $z < x$.

The fact that a computation is non–deterministic or probabilistic does not preclude input and output from being linked via a functional correspondence. Let us consider again the algorithm of Example 2.4. It is easy to realize that it computes the following function (1 indicates a positive answer, 0 a negative answer):

$$f(x) = \begin{cases} 1 & \text{if } x \text{ is not prime} \\ 0 & \text{if } x \text{ is prime} \end{cases}$$

In a similar way there are probabilistic algorithms that with the same input data produce the same output data.

We will reconsider later the concept of non–determinism (see Section 2.5). At present, we are only interested in pointing out that the main open issues of computational complexity are linked to the relationship between determinism and non–determinism.

2.2.2 Computing models

Up to now, we have seen in an abstract way the concepts linked to algorithms. These concepts become more concrete when we consider specific computing models. Two main factors take part in the definition of a computing model: a formal language and a set of computing resources to execute the algorithms written in that language. Some models accurately describe computing resources, while others concentrate on algorithms and avoid details on the machine, considered unimportant. The most widely used computing models are by far the *random access machines* (*RAM*) and the *Turing machines*. A complete description of these computing models can be found in [AHU74].

Random access machines

A random access machine (RAM) is shown schematically in Figure 2.2. A finite–length program controls the computation which in turn uses an infinite set of working registers named r_0, r_1, r_2, ... Two additional registers have a particular meaning and therefore are represented by special symbols:

- PC is the *program counter*, i.e. the register containing at each step of the computation the reference (either the address or the progressive

number) of the next executable instruction;
- A is the *accumulator*, i.e. the register used in each arithmetic instruction and in each transfer of information.

Input data reside in a read–only device (e.g. a tape). Output data are released on a write–only device. Each register, as well as each portion of the read or write devices, may contain an integer. Program instructions may explicitly address any particular working register. Vice versa the accumulator and the program counter are implicitly referred to by program instructions. In a similar way, the current positions of the read and write devices are implicitly referred to by means of pointers (named PI and PO in Figure 2.2) transparent to the program.

Note that the program is not written in memory (as in the case of conventional computers) but is integrated into the machine. In other words, different programs represent different random access machines. Table 2.1 lists the set of instructions of a typical RAM. A *label* is any arbitrary sequence of characters and numbers, starting with a character. Any *operand* is one of the following expressions, where i denotes an integer represented in positional decimal numeration:

- #i (is an *immediate* operand);
- ri (is a *direct* operand);
- @i (is an *indirect* operand).

Labels and operands are the only kind of expression in the language of random access machines.

The meaning of an expression is either an integer or a register. Some expressions have only one meaning. Others denote an integer or a register

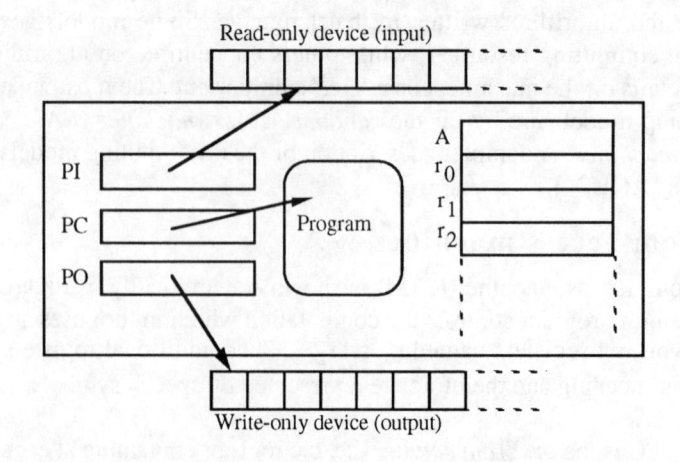

Figure 2.2 Random access machine outline.

Instruction	Meaning
LOAD operand	A ← R(operand)
STORE operand	L(operand) ← R(A)
ADD operand	A ← R(A)+R(operand)
SUB operand	A ← R(A)-R(operand)
JUMP label	PC ← R(label)
JZERO label	R(A)=0 ⇒ PC ← R(label)
JGTZ label	R(A)>0 ⇒ PC ← R(label)
READ	see text
WRITE	see text
HALT	

Table 2.1 RAM instructions.

depending on the context, i.e. the instructions in which they are specified. In Table 2.1, $L(x)$ represents the register specified by the expression x, and $R(y)$ represents the integer specified by the expression y. As a general comment we can make the following remarks:

1. The immediate operand #i always represents an integer value, and more precisely the number i. For example, the instruction LOAD #1 writes the value 1 to the accumulator. For this reason, the instruction STORE #1 is wrong because the expression #1 does not represent any register.
2. The direct operand ri may represent either the integer contained in register r_i or the register r_i itself. For example, in ADD r1, r1 represents the integer contained in r_1. The same expression in STORE r1 represents the register r_1 itself.
3. The indirect operand @ri may also have two meanings, depending on the context. Let us suppose, for example, that register r2 contains the number k. In this case, the expression @r2 contained in LOAD @r2 denotes the integer value contained in r_k. The same expression contained in STORE @r2 denotes the register r_k itself.
4. A label is defined by inserting it before an instruction. The value it gets is the ordinal (positive) number of the related instruction. It is used with this meaning in the JUMP, JZERO and JGTZ instructions. For example, the execution of the instruction JUMP JOHN puts the value contained in the label JOHN into the program counter register. This means that the next executable instruction is the instruction headed by the symbol "JOHN".

14 Preliminaries

Let us now briefly explain the meaning of READ and WRITE instructions as shown in Table 2.1. The execution of a READ instruction causes the transfer of data currently addressed by PI into the accumulator. PI then points to the next data. In a similar way, the execution of a WRITE instruction first causes the updating of the pointer to the next position in the output device, then the transfer to this location of the data contained in the accumulator.

We can now describe in simple terms a RAM computation. n input data are stored in the first n locations of the input device. The address of the first data is written into PI, and n is stored in the accumulator. The remaining registers contain the value 0, except the program counter which contains the address of the first program instruction. The status of a RAM is given by the contents of all registers together with the contents of input and output devices. Each instruction can affect at most one register. It follows that at the kth step of the computation, at most k registers have a value different than zero. For this reason, the status of a RAM computation can always be given by means of a finite number of registers. The order of instruction execution is controlled both by an implicit and by an explicit mechanism. The implicit one is represented by the automatic increment (by 1) of the integer value contained in the PC register. This means that instructions are executed in the same order as they appear in the program. This ordering can only be modified by using jump instructions, i.e. JUMP, JZERO and JGTZ. These instructions modify the content of the program counter, hence determining the next instruction to be executed. Computation halts when (and if) the HALT instruction is executed.

Example 2.5 (multiplication of two integers)
As you may have already noticed, the RAM repertory does not include any multiplication and division instructions. We will expand later on the rationale for this absence. Let us now look at a possible way to perform the multiplication by means of the available RAM instructions. Let x and y be two integers, and let us suppose, for the sake of simplicity, that $0 < x$, and $y < 2^n$.

We can write x in binary format: $x = x_{n-1}2^{n-1} + x_{n-2}2^{n-2} + \ldots + x_0$, where $x_i \in \{0,1\}$, $i = 0, \ldots, n-1$. The following equation holds:

$$xy = x_{n-1}2^{n-1}y + x_{n-2}2^{n-2}y + \ldots + x_0 y.$$

This means that the product can be performed by summing at most n numbers of the form $2^i y$, i.e. those numbers where $x_i = 1$. In turn, each product $2^i y$ can be computed by doubling i times the value y (hence by using only addition). Please note that the worst case requires $2(n-1) + (n-2) + \ldots + 2 + 1$, i.e. $(n^2 + n - 2)/2$ additions.

Figure 2.3 shows the program that computes the product of two integers by using the above algorithm. For simplicity the structure of the program is described in Pidgin ALGOL language. Only the innermost **while** is written in RAM language. Details for a complete RAM code are left as an exercise for the reader.

```
begin
    read r1;                comment r1 contains the number x;
    read r2;                comment r2 contains the number y;
    if r1 = 0 or r2 = 0 then write 0 else begin
        r0 ← 0;             comment r0 stores the result;
        while r1 ≠ 0 do begin
            r4 ← r2;        comment the value of y is doubled at each step;
            r3 ← 1;         comment r3 will contain the max 2^i such that 2^i ≤ x;
            while r3 + r3 ≤ r1 do begin
                r4 ← r4 + r4;
                r3 ← r3 + r3
            end;
            r0 ← r0 + r4;   comment add 2^i y to the result;
            r1 ← r1 − r3;   comment subtract 2^i to x;
        end;
        write r0
    end
end
```

Figure 2.3(a) Compute the product of two positive integers.

```
WHILE2   LOAD    r3
         ADD     r3
         SUB     r1
         JGTZ    END2
         LOAD    r4
         ADD     r4
         STORE   r4
         LOAD    r3
         ADD     r3
         STORE   r3
         JUMP    WHILE2
END2     ...
```

Figure 2.3(b) Portion of the program in Figure 2.3(a) codified in RAM language.

The computation of a function from Z^n to Z can be easily associated to the RAM's set of computations.

Definition 2.6
A random access machine R computes a function f ($f: Z^n \to Z$) if, given $<i_1, i_2, ..., i_n>$ as input, the following conditions hold:

- if $f(i_1, i_2, ..., i_n)$ is defined, the computation of R terminates at the first output value u such that $u = f(i_1, i_2, ..., i_n)$;
- if $f(i_1, i_2, ..., i_n)$ is not defined, the computation of R does not terminate (the HALT instruction is never executed).

The function computed by a RAM R is indicated by f_R.

RAM computations are deterministic. However it is possible to consider a non–deterministic RAM by simply allowing the same label to appear in front of more than one instruction. Consider the following portion of a non–deterministic program

```
            ...
            JUMP AHEAD
            ...
AHEAD       LOAD #1
            ...
AHEAD       LOAD #-1
            ...
```

This program makes sense only if we suppose that the machine itself can make the right choice. In a real machine, the above situation is not allowed because there is no criterion to decide which is the instruction to be executed after JUMP AHEAD.

Simplified RAMs

When developing algorithms, it is often useful to have models which are much more sophisticated than RAMs. As an example, think of numeric problems (requiring little control) and comparison problems (requiring much control). Typical numeric problems have the same execution sequence as the physical ordering of their instructions. Vice versa in comparison problems, such as sorting, different input data sets can determine different execution sequences.

Example 2.7
Let A be a matrix of order n. The elements of A are denoted by $a(i,j)$ or a_{ij}. The type of its elements is a property of the matrix, hence there are real matrices (i.e. matrices whose elements are real numbers), Boolean matrices (i.e. matrices whose elements are Boolean values), and so on. If

multiplication and addition are possible on the elements of two matrices A and B, then the multiplication by rows and columns of A and B is defined. The product is a matrix C, with the generic element c_{ij}, defined by the following equation:

$$c_{ij} = \sum_{k=1}^{n} a_{ik} b_{kj}, \qquad i,j = 1,\ldots,n$$

The algorithm performing the multiplication by rows and columns is a rewriting of the previous definition in a programming language.

```
begin
    for i ← 1 until n do
        for j ← 1 until n do begin
            c(i,j) ← 0;
            for k ← 1 until n do
                c(i,j) ← c(i,j) + a(i,k) * b(k,j)
        end
end
```

Note that the algorithm does not perform any check on input data (except for the indices i, j and k). It follows that having n fixed it is possible to "unroll" the iterations specified by the algorithm, writing the instructions in the same order as they would be executed:

```
c(1,1) ← 0;
c(1,1) ← c(1,1) + a(1,1) * b(1,1);
c(1,1) ← c(1,1) + a(1,2) * b(2,1);
....
c(1,1) ← c(1,1) + a(1,n) * b(n,1);
c(1,2) ← 0;
c(1,2) ← c(1,2) + a(1,1) * b(1,2);
....
c(1,2) ← c(1,2) + a(1,n) * b(n,2);
....
c(n,n) ← 0;
c(n,n) ← c(n,n) + a(n,1) * b(1,n);
....
c(n,n) ← c(n,n) + a(n,n) * b(n,n)
```

The above example is typical of many numeric problems. The fact that no control is needed makes jumping instructions (JUMP, JZERO, JGTZ) useless. Moreover, because the time spent in input/output operations is often much less than the rest of the computation, it is possible to discard READ and

18 Preliminaries

WRITE instructions also. Input data are assumed to be already present in memory before the computation starts, while output data are stored in known variables. Moreover, the following sequence of instructions:

```
LOAD      x
ADD/SUB   y
STORE     z
```

can be replaced by the following assignment:

$$z \leftarrow x \pm y$$

Because the absence of multiplication and division is not acceptable in a model to be used for numeric problems, the following assignments are allowed (see Section 2.4):

$$z \leftarrow x*y \text{ and } z \leftarrow x/y$$

Programs composed only of sequences of assignments are called *straight–line programs*. Note that the assignments in Example 2.7 do not represent a true straight–line program. For example, the assignment:

$$c(1,1) \leftarrow c(1,1) + a(1,1) * b(1,1)$$

includes more than a single operation on the right hand side. This kind of notation is allowed as a simplified form of the following sequence ("temp" is a temporary variable):

$$\text{temp} \leftarrow a(1,1) * b(1,1);$$
$$c(1,1) \leftarrow c(1,1) + \text{temp}.$$

Example 2.8 (Sorting)
Let S be a set on which a linear relation $\leq$ is defined. Consider the problem of sorting on S: for any subset X of S with cardinality n, produce a sequence $x_1, ..., x_n$ such that, if $1 \leq i < j \leq n$, then $x_i, x_j \in X$ and $x_i \leq x_j$. Let us simplify the problem by supposing that the elements to be sorted, $x_1, x_2, ..., x_n$, are all distinct. As each comparison can only have two different results, the behaviour of any algorithm can be described by a binary tree in which any node represents a comparison, and the arcs originating from it represent the possible results (see Figure 2.4). A leaf of the tree represents a possible solution to the problem.

An appropriate model for comparison problems is the representation of programs by a *decision tree*. We will later see how this model is useful for complexity studies. A decision tree is an abstract view of a RAM program

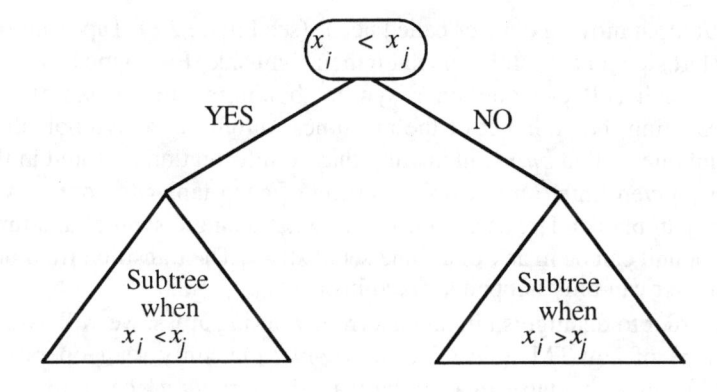

Figure 2.4 Tree describing a sorting algorithm.

which specifies only control information.

Turing machines

Turing machines (TM) are the most primitive computing model. They were introduced by Alan Turing as the formal characterization of the algorithm and computability concepts.

A TM is composed of a control unit and a set of k tapes with relative

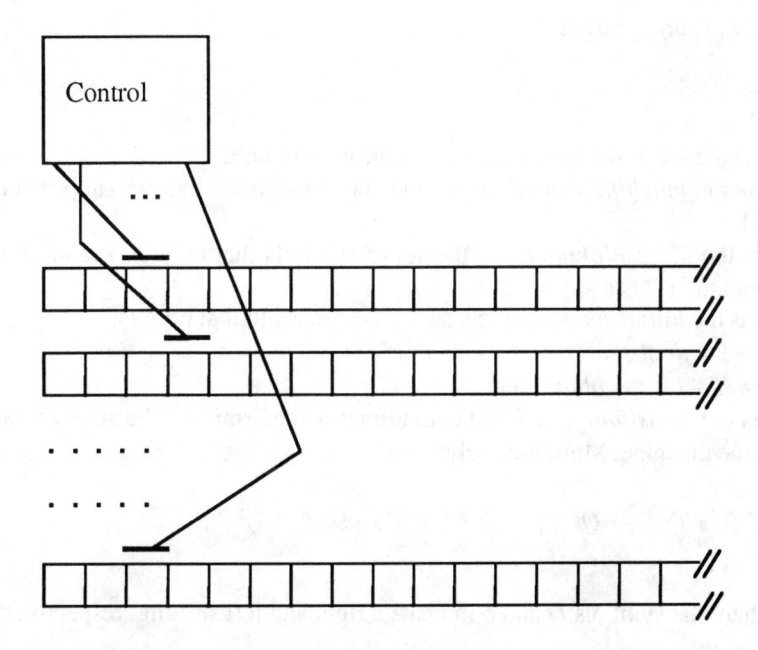

Figure 2.5 TM outline

input/output moving devices called heads (see Figure 2.5). Tapes are finite on their left side, but are unbounded on their right side. Each tape is divided into cells. Each cell can contain a symbol belonging to the set of symbols representing the alphabet of the machine. Among these symbols there is a special one, called *blank*, indicating that no information is stored in the cell. At any given time, only a finite portion of each tape can contain symbols other than blanks. Heads can only read/write a single symbol at a time. The control unit can be in any of a finite set of states. The transition from one state to another can only happen at fixed instants $t_0, t_1, t_2, \ldots$.

In order to distinguish input data from working areas, we will assume that one tape of any TM is used as the *input tape*: only read operations are possible on such a tape. In a similar way, the *output tape* can only be used to perform write operations. The remaining $k - 2$ ($k \geq 3$) tapes are called *working tapes*: on each of them both read and write operations are possible. In the following, we will assume that the input tape and the output tape are the first and the last tape, respectively.

A step of a TM computation consists of the following operations: (1) the control unit changes its state as a consequence of its current state and of the symbols read by the heads; (2) each head (except for the input tape head) writes a new symbol in the relative current cell; (3) each head moves a cell right or a cell left or does not move at all.

Definition 2.9
A TM with k tapes is a 7–tuple:

$(Q, T, I, q_0, q_f, b, \delta)$

where:

Q is the set of the possible *states* of the control unit;
T is the *machine alphabet*, i.e. the set of symbols that may appear on a cell;
I is the *input alphabet*, i.e. the set of symbols that may be found on the input tape (I is a subset of T);
q_0 is the *initial state*, i.e. the state of the control unit at time t_0;
q_f is the *final state*;
$b \in T - I$ is the *blank* character;
δ is the *transition function*, i.e. the function determining the state changes in the machine. More precisely:

$\delta : Q \times T^{k-1} \to Q \times \{D, S, F\} \times (T \times \{D, S, F\})^{k-1}$

where the symbols D and S indicate a right and left shifting, respectively.

The symbol F indicates that the head is steady.

The transition function of a TM can be viewed as an algorithm. Any language used to describe functions can be used to express this algorithm (see Example 2.14 where a table is used).

At any given time, the machine state (called hereafter *configuration*) is determined by the state of the control unit, the contents of the tapes, and the head positions. A formal description of a TM configuration at a particular time is called an *instantaneous description*.

Definition 2.10
An instantaneous description (ID) is a k–tuple $(\alpha_1, \alpha_2, ..., \alpha_k)$, where each α_i is a string representing tape i. More precisely, each α_i can be written as $\beta_i q z_i \gamma_i$, where:

$\beta_i \in T^*$ ($\beta_1 \in I^*$) is the sequence of symbols which is present on the portion of tape i starting from the left end up to the head position;
q is the current state of the control unit;
$z_i \in T$ ($z_1 \in I$) is the symbol just read from head i;
$\gamma_i \in T^*$ ($\gamma_1 \in I^*$) is the sequence of symbols which is present on the portion of tape i starting from the position on the right of the head to the last symbol that is not a blank at the right of the head.

The starting configuration of a TM is described by the following starting ID: $(q_0 z_1 z_2 ... z_n, q_0, ..., q_0)$, where $z_1 z_2 ... z_n$ is the input sequence stored on tape 1. This means that at the beginning the control unit is in state q_0, all the heads are positioned on the first cell of the relative tape, input data is on the first tape, and the other tapes only contain blanks. An ID $(\alpha_1, \alpha_2, ..., \alpha_k)$ is final if each α_i can be written as $\beta_i q_f z_i \gamma_i$.

Definition 2.11
A TM computation is a sequence of instant descriptions $ID_0, ID_1, ID_2, ...,$ such that ID_0 is the starting description, and ID_{i+1} results by applying function δ to ID_i (written $ID_i \vdash ID_{i+1}$).

If there exists an $n \geq 1$ such that:

$$ID_0 \vdash ID_1 \vdash ... \vdash ID_n$$

and ID_n is a final description, then the computation converges. It diverges otherwise. It is possible to associate with the computations of a TM, the computation of a function from I^* to $(T - \{b\})^*$. To this end, we can consider, as the result of a converging computation, the string produced on the output tape starting at the leftmost cell (included) up to the first cell containing the blank character.

Definition 2.12
A Turing machine M computes the function f ($f: I^* \to (T - \{b\})^*$) if, on an input $z \in I^*$, the following conditions are verified:

- if f is defined on z, then there exists an n such that $ID_0 \vdash ID_1 \vdash \ldots \vdash ID_n$, where ID_n is the final instant description; moreover, the string $w \in (T - \{b\})^*$ produced by M on the output tape is such that $f(z) = w$;
- if $f(z)$ is not defined, the computation diverges.

Turing machines can then be considered as devices for manipulating strings.

However, by considering particular codings, it is possible to associate with a TM the computation of functions on several domains. As an example, let us consider how a TM can compute a function defined on an n–tuple of natural numbers. In order to represent an n–tuple of natural numbers, it is possible to think of an alphabet composed of four different symbols, $T = \{0, 1, \#, b\}$, where "0" and "1" are used to represent the natural numbers in binary format, "#" is used to separate the elements of the n–tuple, and b is the blank character. If $i(z)$ is a positive integer number the binary representation of which is the string z, we can represent the n–tuple $<i(z_1), i(z_2), \ldots, i(z_n)>$ with the string $z_1 \# z_2 \# \ldots \# z_n$.

Definition 2.13
A Turing machine M computes the function f ($f: N^n \to N$) if, on an input $z_1 \# z_2 \# \ldots \# z_n$, the following conditions are verified:

- if f is defined on $<i(z_1), i(z_2), \ldots, i(z_n)>$, then there exists an m such that $ID_0 \vdash ID_1 \vdash \ldots \vdash ID_m$, where ID_m is the final instant description; moreover, the string w produced by M on the output tape is such that $i(w) = f(i(z_1), i(z_2), \ldots, i(z_n))$;
- if $f(i(z_1), i(z_2), \ldots, i(z_n))$ is not defined, the computation diverges.

The function computed by a Turing machine M is indicated with f_M.

Example 2.14 (design of a TM computing the sum of two binary integers)

The machine uses three tapes. The first tape contains the input string $z_1 \# z_2$, where $i(z_1)$ and $i(z_2)$ are the numbers to be added. For simplicity, we suppose that the numbers are represented starting from the least significant bit. Figure 2.6 shows the table of the transition function. At the beginning (i.e. state 0) the TM marks the first cell of tape 2 (working tape) with the special symbol "#".

At state 1 the portion of tape 2 between the first character (i.e. the character in the leftmost cell) and the character "#" (not included) is copied to

Basic computability theory 23

Status	Symbol on the tape 1 2 3	Symbol written, Head shift 1 2 3	New status
0	0 b b	0,F #,D b,F	1
0	1 b b	1,F #,D b,F	1
1	0 b b	0,D 0,D b,F	1
1	1 b b	1,D 1,D b,F	1
1	# b b	#,D b,S b,F	2
2	0 0 b	0,F 0,S b,F	2
2	0 1 b	0,F 1,S b,F	2
2	1 0 b	1,F 0,S b,F	2
2	1 1 b	1,F 1,S b,F	2
2	0 # b	0,F #,D b,F	3
2	1 # b	1,F #,D b,F	3
3	0 0 b	0,D 0,D 0,D	3
3	0 b b	0,D b,F 0,D	3
3	b 0 b	b,F 0,D 0,D	3
3	1 0 b	1,D 0,D 1,D	3
3	1 b b	1,D b,F 1,D	3
3	0 1 b	0,D 1,D 1,D	3
3	b 1 b	b,F 1,D 1,D	3
3	1 1 b	1,D 1,D 0,D	4
3	b b b	b,F b,F b,F	5
4	0 0 b	0,D 0,D 1,D	3
4	0 b b	0,D b,F 1,D	3
4	b 0 b	b,F 0,D 1,D	3
4	1 0 b	1,D 0,D 0,D	4
4	1 b b	1,D b,F 0,D	4
4	0 1 b	0,D 1,D 0,D	4
4	b 1 b	b,F 1,D 0,D	4
4	1 1 b	1,D 1,D 1,D	4
4	b b b	b,F b,F 1,D	5
5			

Figure 2.6 Transition function for the TM of Example 2.14.

the working tape. At state 2 the head of tape 2 is moved to the first bit of the copied number; the special character positioned at the beginning of the working tape is used for this operation. At states 3 and 4 the bits of the two numbers (on tape 1 and 2) are read and added, and the result is written to tape 3 (output tape). At state 4 the machine keeps track of the amount to be carried over from the preceding position. Note that the program shown in Figure 2.6 still works for numbers with different lengths.

Example 2.15
The computation generated by the TM of Example 2.14 on the input "01#11" is shown in Figure 2.7.

24   Preliminaries

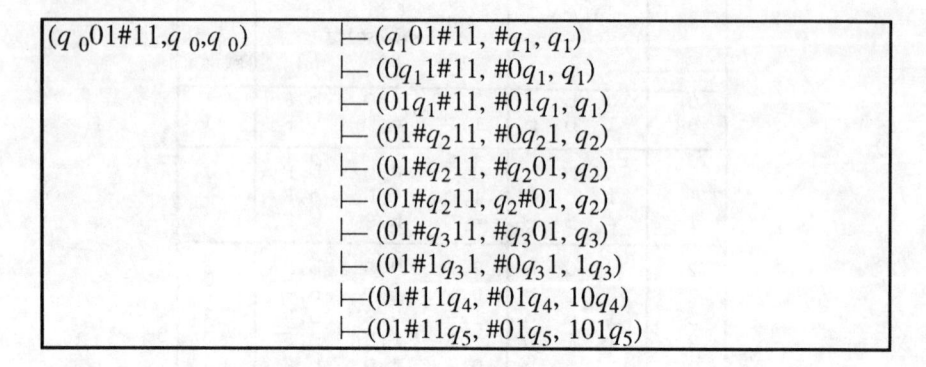

```
(q_0 01#11, q_0, q_0)    ⊢ (q_1 01#11, #q_1, q_1)
                         ⊢ (0q_1 1#11, #0q_1, q_1)
                         ⊢ (01q_1 #11, #01q_1, q_1)
                         ⊢ (01#q_2 11, #0q_2 1, q_2)
                         ⊢ (01#q_2 11, #q_2 01, q_2)
                         ⊢ (01#q_2 11, q_2 #01, q_2)
                         ⊢ (01#q_3 11, #q_3 01, q_3)
                         ⊢ (01#1q_3 1, #0q_3 1, 1q_3)
                         ⊢ (01#11q_4, #01q_4, 10q_4)
                         ⊢ (01#11q_5, #01q_5, 101q_5)
```

Figure 2.7 A computation of the TM shown in Example 2.14.

As in the case of random access machines, the concept of non–determinism can be defined on Turing machines. In a non–deterministic TM the transition function maps elements of $Q \times T^{k-1}$ into subsets of $Q \times \{D, S, F\} \times (T \times \{D, S, F\})^{k-1}$ (instead of single elements). In practice the transition function produces a finite set of possible next configurations.

2.2.3 Computability and complexity

A precise definition of the algorithm concept was one of the main objectives of *computability theory*. The need for such a definition, in a time when computers did not yet exist,[2] was due to the studies on fundamentals of mathematics. The coming of computers brought a new and strong impulse to these studies, though their basis had already been developed.

In order to understand reasons and objectives of concrete computational complexity, which is the main subject of this book, we first want to list the main results of classic computability theory:

- A precise statement of the algorithm concept has been developed through the definition of formal computing models, such as Turing machines.
- It has been proven that all proposed models can describe algorithms able to compute the same set of functions. A false problem is the use of different kind of data. As an example, Definition 2.6 states that the function computed by a RAM has its domain in Z^n, while Definition 2.12 states that the function computed by a Turing machine has its domain in a set of strings. However, Example 2.14 proves that Turing machines can also compute, by using an appropriate coding, functions defined on natural numbers. To realize

---

[2] The first most significant computability studies are from the thirties, executed by Church, Godel, Kleene and Turing.

the fact that this is actually a false problem, just consider that all input and output information is actually represented in a computer by sequences of symbols on a given alphabet.
- Model equivalence brought Church and Turing to formulate the following statement: "If a function can be computed by an ordinary computing model, then it can be computed by a Turing machine". The great importance of this statement (called the *Church–Turing thesis*) is that it applies the "computability" concept to functions. In other words, it allows functions to be classified according to their computability (a computable function is also called *recursive*), no matter which computing model is used.

  An important recursive function is the program *interpretation function*. If f_p denotes the function computed by program p, the interpretation function g is defined as $g(p, x) = f_p(x)$. The fact that g is computable means that there exists a Turing machine U such that, receiving the input pair $<d(M), x>$, where $d(M)$ is the description of any Turing machine M and x is input to M, it simulates the computation performed by M on the input x. If this computation converges, then the computation performed by U on the input $<d(M), x>$ converges and produces the same result; if the computation diverges, then the computation performed by U also diverges. U is called the *universal Turing machine*. Note that the description of M is mainly a coding of the transition function to a string of a given alphabet.
- It has been proven that some functions are not computable. A key problem to which it is not possible to associate a computable function is the program termination problem, well known in computer science: "Given a program expressed in any computing model, with certain input data, determine whether the computation originated by the program on the input data converges".

Classic computability clearly separates computable functions from non–computable ones. Computational complexity, originating from classic computability, concentrates on evaluating the amount of work (i.e. resources, time and memory) required to compute classes of recursive functions. The working hypotheses are often different in the two fields:

1. Having unlimited resources (e.g. an infinite tape) is crucial for an abstract characterization of computable functions. On the contrary computational complexity considers resources as a function of input size.
2. Each sensible definition of computability states that programs must be uniform, i.e. definable with a finite quantity of information. Interesting applications in the study of complexity can be found if we release this constraint. As an example (see Section 3.3 on

Boolean circuits), think of the studies on the minimum quantity of resources needed to solve a given problem.

3. Computability theory firmly states that the possibility of non–termination in a computation is intrinsic to algorithms. For example, we know that if we accept the Church–Turing thesis, then any model computing only total functions cannot be complete (i.e. it cannot compute all total functions).

4. In classic computability, where computations have infinite length, non–determinism is a minor factor. A deterministic machine can always simulate a non–deterministic machine by trying all possible alternatives in a systematic way. In computational complexity, on the contrary, determinism plays an important role in expressing key properties of complexity classes, and in identifying some key issues (e.g. the question $P =? NP$).

5. Probabilistic devices, excluded from classic computability, play an important role in solving problems with limited computational costs.

Many approaches are possible to computational complexity [BM75]. Among them is the approach linked to computation theory which tries to characterize the computational complexity of functions independently of any particular model. This approach refers to the general concept of computing *resources* (e.g. computing time, memory, number of performed tests, maximum value contained in registers during computation, etc.). This *abstract* approach mainly provides general complexity limitations. A major result is the proof that there are computable functions for which there is no optimum algorithm, i.e. there is no algorithm employing the minimum quantity of needed resources (see Section 2.4).

Another approach considers specific computing models (e.g. Turing machines) and particular complexity parameters (e.g. time and memory). It characterizes the complexity of functions on the basis of the amount of resources required for the computation. A problem linked to this concrete approach is the dependence from the selected model. However, there are some important examples (e.g. NP-complete problems) where the computing model has no influence.

A further approach basically looks for efficient algorithms of important function classes. Here the computing model is generally an ideal computer and its details are not considered to be important.

This book considers only the last two approaches, which will be referred to as *concrete complexity* approaches. In simple terms, we can say that the main objective of concrete complexity approaches is the characterization of what is efficiently computable, and the classification of algorithms, based on their use of resources, and problems, based on their implicit *demand* for resources.

2.3 Functions, problems, languages

This section formally introduces the concepts of "problem" and of "algorithm solving a problem". So far we have considered algorithms as means for computing functions, following the ideas of computability theory. We will now show how the concept of problem is a more general one. Both the function and language "recognition" concepts can lead again to the problem concept.

A problem R consists of a set D_R and a collection of sets $C_R(d)$, one for each $d \in D_R$. The elements of D_R are called *instances* of the problem R. The elements of the set $C_R(d)$, corresponding to instance d, are called *solutions* of d. Hence a problem is defined in terms of its instances (or alternatively in terms of what are considered to be its instances).

Example 2.16
Let us consider the following problem: "solution of second order equations with integer coefficients on R", indicated in the following with EQSOL. This problem consists of the set D_{EQSOL} containing all equations like $ax^2 + bx + c = 0$, where $a, b, c \in Z$, and of the collection of subsets of R $<C_{EQSOL}(d)>_{d \in D_{EQSOL}}$. The elements belonging to $C_{EQSOL}(d)$ are the solutions of equation d. $C_{EQSOL}(d)$ does not contain more than two elements. As an example, $C_{EQSOL}(x^2 + 1 = 0) = \emptyset$, and $C_{EQSOL}(x^2 - 1 = 0) = \{-1, 1\}$.

From the previous example and from the definition of a function (see Section 2.2.1) it follows that a problem is a rule associating elements of a set (of solutions) to elements of a set (of instances). Exactly like functions. The only difference is that a certain instance of a problem may have more than one solution. In this sense, functions are a particular class of problem.

The problem concept can be formalized by the notion of relationship between sets. Note that a relationship[3] between sets D and C is a subset M of the Cartesian product $D \times C$ (the set of all possible pairs (d, c), such that $d \in D$ and $c \in C$). If each element $d \in D$ is present at most one time as first element of a pair of M, then M is a function. The sets D and C can be very different. They can even be created from other sets. As an example, EQSOL can be considered as a relationship between sets Z^3 and R.

An algorithm (in a given computing model) solves problem R if, given as input a description of an instance d belonging to R, when $C_R(d) \neq \emptyset$ it returns a solution of d. Otherwise it returns a special symbol, such as the Boolean FALSE.

Before introducing the formal definition of problem, we shall consider more complex examples. For this purpose let us first introduce the notions of graph and directed graph.

---

[3]To be more precise we should have used the term correspondence, because a relationship is a correspondence between a set and itself (i.e. it is a particular case of correspondence).

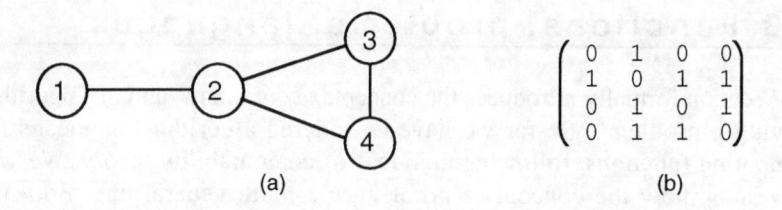

Figure 2.8 Representations of a graph.

Definition 2.17
A *graph* G is a pair $G = (V, E)$, where V is a finite set and E is a set consisting of unordered pairs of elements belonging to V. The elements of V are called *vertices* or *nodes* of the graph; E elements are called *edges*. Since the type of elements of V is not important, we will always indicate a graph with n nodes with the notation $V = \{1, 2, ..., n\}$. A *path* with length k on a graph is a sequence of $k+1$ nodes $c = (i_1, i_2, ..., i_k, i_{k+1})$, such that $(i_j, i_{j+1}) \in E, j = 1, ..., k$. Moreover the graph is called *labelled* if a function $\lambda: V \to \Omega$ is defined on V, where Ω is any set (its elements are called labels).

A graph $G = (V, E)$ can be graphically represented by using circles to represent nodes, and lines to link the circles corresponding to pairs of nodes in E. If the graph is labelled, labels are written inside the circles of the corresponding node. If no explicit labelling is used, we assume $\lambda: V \to \{1, 2, ..., n\}$ (see Figure 2.8(a)). Another way to represent the graph, and a useful one for computer manipulation, is the adjacency matrix. The adjacency matrix $A_G = (a_{ij})$ of a graph G with n nodes is such that $a_{ij} = a_{ji} = 1$ if $(i, j) \in E$, 0 otherwise (see Figure 2.8(b)). Hence the adjacency matrix of a graph is a symmetric matrix.

For many problems, it is important to assign a direction to the edges of a graph. The result is a directed graph.

Definition 2.18
A *directed graph* (or *digraph*) G is a graph, i.e. $G = (V, E)$, in which the elements belonging to E are ordered pairs. In other words, $E \subseteq V \times V$ is a relation on V. Edges in a digraph are usually called *arcs*. A node i is an *immediate predecessor* of node j if the arc (i, j) exists. A *path* of length k on a digraph is a sequence of k arcs $(i_1, i_2), (i_2, i_3) ... (i_k, i_{k+1})$, where all $i_j, j = 1, ..., k$, are not necessarily distinct. A *loop* in G is a path $(i_1, i_2), (i_2, i_3) ... (i_k, i_{k+1})$ such that $i_1 = i_{k+1}$. If G has no loops, it is called an *acyclic graph*.

The graphic representation of a digraph uses arcs instead of lines. In general, the adjacency matrix is not symmetric (see Figure 2.9).

In Chapters 4, 5, and 6 we will use another structure, called a *tree*, which is a particular case of a graph. A (direct) tree is a graph having the following properties:

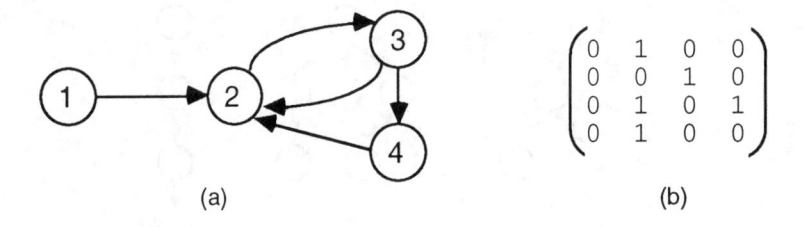

Figure 2.9 Representations of a digraph.

1. there is only one node, called the *root*, with no immediate predecessors;
2. every node except for the root has only one predecessor;
3. for each node i (different than the root) there exists a path joining the root to i.

A tree where each node is an immediate predecessor of at most two nodes is called a *binary tree*. A set of trees is called a *forest*. It is possible to obtain a tree from a forest by simply adding a root node (see Figure 2.10).

Many important problems can be represented by means of graphs.

Definition 2.19 (Hamilton direct path)
Let $G = (V, E)$ be a digraph with n nodes. A *Hamilton path* on G is a path with length n, reaching all nodes (i.e. a sequence of arcs (i_1, i_2), (i_2, i_3) ... (i_{n-1}, i_n) where all i_j are distinct).

Definition 2.20 (k–clique)
Let $G = (V, E)$ be a graph with n nodes. A *clique* with dimension k ($k \leq n$), or k–clique, is a subset V' of V with k elements completely connected, i.e. such that for each $i, j \in V'$, then $(i, j) \in E$. As an example the subgraph composed by nodes 2, 3 and 4 of Figure 2.8 is a 3–clique.

Let us now consider the following problems: (1) determine a Hamilton path in a digraph; (2) determine a k–clique in a graph. Hereafter we will indicate these problems with HAMILTON and CLIQUE(k) respectively.

Let us now consider again the main issue of formalizing the problem concept. We can associate with each instance of a problem a number, called the *size* of the instance, dimensioning the instance itself. When dealing with problems on a graph, a natural choice is the number of nodes.

Definition 2.21
Let R be a problem. R_n denotes the problem composed of all instances of R with size n, based on a given measure. We denote this concept with the

30 Preliminaries

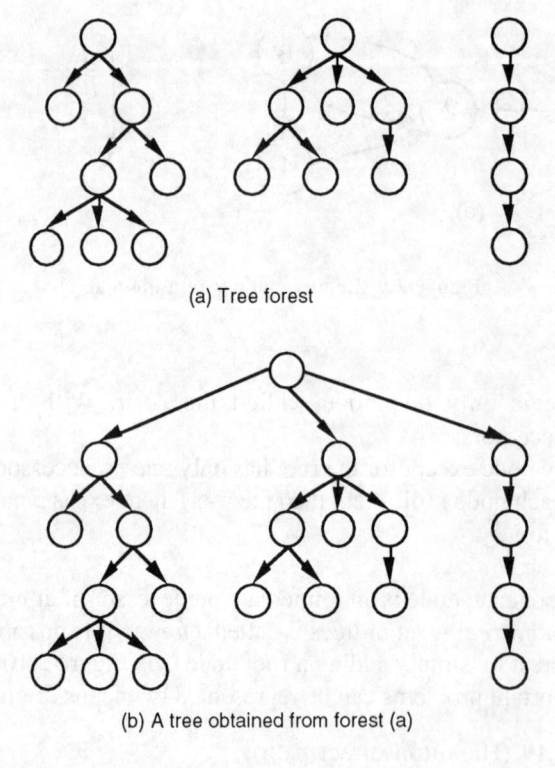

(a) Tree forest

(b) A tree obtained from forest (a)

Figure 2.10 Trees and forests.

notation $R = \langle R_n \rangle_{n \in N}$.

Let us consider the following examples:

- HAMILTON = $\langle \text{HAMILTON}_n \rangle_{n \in N}$;
- CLIQUE(k) = $\langle \text{CLIQUE}_n(k) \rangle_{n \in N}$;

where HAMILTON$_n$ (CLIQUE$_n(k)$) is the set of instances of HAMILTON (CLIQUE(k)) defined on a graph with n nodes, when the size is given by the number of nodes. This approach permits us to define a problem by means of a fixed–length problem.

Example 2.22 (HAMILTON$_n$ description)
Let us represent a (directed) graph by means of its adjacency matrix. Let n be the size representing the number of nodes (i.e. the matrix order). A graph description does not require more than n^2 bits. A HAMILTON$_n$ instance can then be formally described on the domain $\{0, 1\}^{n^2}$. A solution to a HAMILTON$_n$ instance is a permutation of the set $\{1, ..., n\}$ which describes either the path found or the symbol NO, indicating that the graph has no

Hamilton path. Formally, a solution to a HAMILTON$_n$ instance is described on the domain $\Pi_n \cup \{NO\}$, where Π_n denotes the set of permutations defined on $\{1, ..., n\}$. These concepts can be summarized with the following notation:

$$\text{HAMILTON}_n \subseteq \{0,1\}^{n^2} \times (\Pi_n \cup \{NO\})$$

Example 2.23 (CLIQUE$_n(k)$ description)
Let us represent a graph by means of its adjacency matrix. Let n be the size representing the number of nodes. The matrix is symmetric and its elements on the diagonal are equal to zero. It follows that in order to describe an instance it is sufficient to consider that part of the matrix above (or below) the diagonal itself. A description of the graph does not require more than $n(n-1)/2$ bits. Because $k \in \{1, ..., n\}$, a CLIQUE$_n(k)$ instance can then be formally described on the domain $\{0,1\}^{n(n-1)/2} \times \{1, ..., n\}$. A solution to a CLIQUE$_n(k)$ instance is either a k-tuple of numbers on the set $\{1, ..., n\}$ or the symbol NO. It follows that:

$$\text{CLIQUE}_n(k) \subseteq \left(\{0,1\}^{n(n-1)/2} \times \{1, 2, ..., n\}\right) \times \left(\{1, 2, ..., n\}^k \times \{NO\}\right)$$

We can now give the formal description of a problem.

Definition 2.24
Let $g_i, h_i : N \to N$ be monotonic non-decreasing functions, $i = 1, 2, ...$. A problem R is a family of relations $R = \langle R_n \rangle_{n \in N}$, such that each instance in R_n has size n and it is:

$$R_n \subseteq \left(D_1^{g_1(n)} \times D_2^{g_2(n)} \times ... \times D_r^{g_r(n)}\right) \times \left(C_1^{h_1(n)} \times C_2^{h_2(n)} \times ... \times C_s^{h_s(n)}\right)$$

where D_i and C_j are well-defined sets, $i = 1, ..., r; j = 1, ..., s$.

Example 2.25 (sorting)
Let us consider the sorting problem (see Example 2.8), which we will indicate with ORD. A natural choice for the size of an ORD instance is the number n of elements that have to be sorted. If the elements belong to a set S (linearly ordered), an ORD$_n$ instance can be formally represented with an n-tuple of elements belonging to S. Analogously for an ORD$_n$ solution. Hence ORD$_n \subseteq S^n \times S^n$ and ORD $= \langle$ORD$_n\rangle_{n \in N}$. It follows that (see Definition 2.24): $r = s = 1$, $D_1 = C_1 = S$, $g_1(n) = h_1(n) = n$.

Definition 2.26
A deterministic algorithm can solve a problem $R = \langle R_n \rangle_{n \in N}$ (where R_n is as given in Definition 2.24) if and only if it computes a family of functions $f = \langle f_n \rangle_{n \in N}$, such that:

$$f_n : D_1^{g_1(n)} \times D_2^{g_2(n)} \times \ldots \times D_r^{g_r(n)} \to C_1^{h_1(n)} \times C_2^{h_2(n)} \times \ldots \times C_s^{h_s(n)}$$

and the following conditions hold: on input X belonging to the domain of f_n, such that X describes an instance of R with at least one solution and with size n, the algorithm produces the output $Y = f_n(X)$ such that $(X, Y) \in R_n$. In this case we say that the function f *realizes* problem R.

Note that two algorithms computing different functions may solve the same problem, because a problem may have more than one solution.

Of particular interest are problems defined on $D \times \{\text{YES}, \text{NOT}\}$, for some D. They are called *decision problems* because their solution depends on the choice between two alternative values (YES and NOT). Language recognition problems are an example of these problems.

Definition 2.27
Let A be a language on alphabet Σ (i.e. $A \subseteq \Sigma^*$). An algorithm recognizes (or decides) A if, on an input $s \in \Sigma^*$, it returns YES if $s \in A$, and NOT otherwise.

A decision problem can be considered as a language recognition problem. In order to do this it is necessary to define the recognized language as the set of instances for which the algorithm returns YES. Obviously, the instances must be codified as strings on an alphabet.

Example 2.28
Let F be a field. Let us consider the problem, indicated with DET_n, of verifying if a matrix $A \in F^{n \times n}$ is non–singular (i.e. its determinant is not null). Each instance of DET_n can be described on the domain $F^{n \times n}$. If it is possible to codify the elements of F as strings on a finite alphabet Σ, then DET_n can be considered as the problem of recognizing the language containing all and only the strings on Σ codifying non–singular matrices on $F^{n \times n}$. A numeric field satisfying the above property is the field of rational numbers.

Example 2.29 (DCLIQUE(k))
Let us consider the graph $G = (V, E)$. The problem of determining if there exists a k–clique in G will be denoted with DCLIQUE(k). This problem can be considered as a language recognition problem. The legal strings of the language are represented by the decimal representation of the size k of the

clique and by a sequence giving the arcs of the graph. If we denote d_i the decimal representation of vertex v_i, it follows:

$$k\,(d_{j_1}, d_{j_2})\,(d_{j_3}, d_{j_4})\,\ldots\,(d_{j_r}, d_{j_{r+1}})$$

where each j_s is not necessarily distinct from j_t, $1 \leq s, t \leq r + 1$. If one considers the following graph:

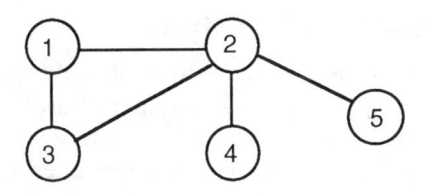

the problem of determining if a 3-clique exists is equivalent to recognizing the string "3(1,2)(1,3)(2,3)(2,4)(2,5)".

Example 2.30 (graph accessibility problem)
Given a digraph G and two distinct nodes x and y, the problem of deciding whether there exists a path linking x to y in G is called the graph accessibility problem (GAP). A GAP instance can be described by listing first the arcs of a digraph, and secondly the pair of nodes to be connected. As an example the string:

"(1,2)(2,3)(3,2)(3,4)(4,2)1,4"

describes the GAP instance for the digraph in Figure 2.9 where the nodes to be connected are numbered 1 and 4. The string belongs to the language defined by GAP because the digraph has a path linking node 1 and node 4.

2.4 Concrete computational complexity

In the previous sections we have summarized the concepts of algorithm, problem and computation model. Both classic computability and computational complexity are based on these ideas. Concrete computational complexity includes three closely interconnected main areas:

- algorithm synthesis;
- problem complexity;
- relationships among computation models.

Algorithm synthesis concentrates on the search for efficient algorithms with respect to computing resources. *Problem complexity* studies the "difficulty" inherent to a problem with respect to its algorithmic solution.[4] These two areas refer to particular computation models either explicitly or implicitly. The third area seeks to single out the existing *relationships among computation models*. The existence of such relationships allows the complexity results proved on a particular computing system to be transferred to different models.

2.4.1 Algorithms: costs criteria

We have already shown in the description of the conventional computer structure that the key computing resources, at least for sequential computations, are time and storage space. The cost of an algorithm is determined by the cost of the computations it generates. For this reason let us first consider the criteria for determining the cost of a computation. Let $C(M)$ denote the set of all possible computations of a machine M. In general the cost in time, T, and the cost in space, S, are functions from $C(M)$ to $R \cup \{\bot\}$, where $\bot$ denotes an indefinite value.

The first objective is to establish a unit measure for space and time. Note that a computation is a sequence of states of the machine. The transition from state to state is originated by the execution of a control program instruction. A possible time unit measure consists of "counting" the number of state transitions if we assume that the time needed to execute an instruction is the same for all instructions and is independent of its state. This is called the *unit cost* hypothesis.

Definition 2.31
The *length* $l(C)$ of a converging computation:

$$C = S^{(0)}, S^{(1)}, ..., S^{(k)}$$

is the number of state transitions, i.e. $l(C) = k$.

Definition 2.32
If the unit cost hypothesis is valid, the time cost for a computation C is defined in the following way:

$$T(C) = \begin{cases} l(C) & \text{if } C \text{ is convergent} \\ \bot & \text{otherwise} \end{cases}$$

---

[4]The term complexity in this book is only used in relation to problems (and computing models), and not to algorithms. When dealing with algorithms we use the term cost, which measures the usage of computing resources.

A unit cost hypothesis exists for space too. Note that a possible measure of the space needed to store a single "datum" is the length of its representation as a string on a given alphabet. The unit cost hypothesis considers the case of data representations with comparable lengths. As an example think of a data representation with a fixed maximum length (usually dependent on the size of the problem). In this way it is possible to think of any datum as occupying a "memory unit".

Definition 2.33
If the unit cost hypothesis is valid, the space cost $S(C)$ of a converging computation C is the total number of different memory locations (e.g. RAM registers or TM working tape cells) used during the computation. If C is diverging, $S(C)$ is undefined.

As far as space is concerned, the cost of a TM can be considered as the maximum number of used cells on each working tape, and not the sum of these cells. This is because the number k of tapes is finite, and the number obtained by summing the cells in the latter case is greater only by a constant factor (Section 2.4.2 will show that a constant factor does not influence asymptotic complexity evaluations).

The cost measures described in Definitions 2.32 and 2.33 are very simple. However the hypotheses on which they are based are not always valid, depending on the computing model and on the computation itself.

In the case of a TM, the unit cost hypotheses are valid. In fact, the machine always executes the same "instruction" (i.e.: (1) change status, (2) write and (3) move heads). Moreover, each single cell may contain only a symbol belonging to a finite alphabet. Vice versa, in the case of conventional computers as well as of RAMs, the unit cost hypotheses are valid only for certain computations.

Let us consider the case of RAMs for which we know the instruction set. Some arguments are against the above hypotheses. Intuitively, indirect operand instructions take more time than direct operand instructions because their execution needs to read from memory twice. Another useful comparison is between instructions which are executed inside the machine and read/write instructions. The latter obviously require more time. This kind of problem can be easily solved in the great majority of cases. Read/write instructions can usually be neglected when dealing with intensive (i.e. CPU bound) computing problems. Moreover, they can be considered extraneous to the real algorithmic part. The issue of the direct/indirect operand can be solved by assuming that an indirect operand instruction takes at most twice the time required by the equivalent direct operand instruction. Hence a computation originated by the execution of indirect operand instructions costs at most twice the cost of the execution of direct operand instructions. We will show in the following that a constant increment of the cost is not a major point against the adoption of the unit cost hypothesis.

36 Preliminaries

The real problem is the magnitude of the values used in the computation. It is unrealistic to impose that the cost be independent of the magnitude of the treated value. A more realistic approach assumes the cost to be a function of the length of the representation. This assumption leads to the definition of the *logarithmic cost* measure, used both for time and space. We use the term "logarithmic cost" because the length of the representation of an integer number can be approximated with the logarithm, in the base of the representation, of its absolute value. More precisely, the number of required digits for the base 2 representation of the signed integer number i is:

$$L(i) = \lceil \log(|i| + 1) \rceil + 1$$

Using the logarithmic cost measure, the cost of a RAM instruction is computed considering all the integer values, including addresses, which are part of the execution of an instruction.

Example 2.34
Let us consider the time cost of the RAM instruction "ADD ri" with respect to the logarithmic measure. The numbers involved in the computation are: (a) the value contained in the accumulator, (b) the address i, (c) the value contained in register r_i. Let $c(i)$ indicate the value contained in register r_i and $c(A)$ the value contained in the accumulator. The estimated cost T_{ADD-D} (ADD–D indicates that it is an "addition with direct operand") of the instruction is the sum of $L(c(A))$, $L(i)$ and $L(c(i))$. The cost of the other RAM instructions can be computed in a similar way.

In order to apply the logarithmic cost measure to space, we assume that the cost assigned to each register r_i used in the computation is $\max\{L(c(i))\}$, where the maximum is computed on all values contained in r_i during the computation.

Definition 2.35
Using the logarithmic cost measure, the time cost $T_l(C)$ (or simply $T(C)$, when the measure is clear from the context) of a converging computation C is the sum of the costs of the single executed instructions. The cost is undefined if the computation diverges.

Definition 2.36
Using the logarithmic cost measure, the space cost $S_l(C)$ (or simply $S(C)$) of a converging computation C is:

$$S_l(C) = \sum_{i \in I} \max\{L(c(i))\}$$

where I is the set of registers used during the computation. If the computation diverges, then $S_l(C)$ is undefined.

The unit cost hypothesis gives an optimistic measure. On the other hand, the logarithmic cost hypothesis sometimes provides a pessimistic evaluation. Let us consider the case of a conventional computer. If every number can be stored in a memory location, the hypothesis that the time required to access and compute a number is independent from its value can be considered to be correct. Moreover (see Sections 3.3 and 4.2), it is possible to implement the algorithms for the basic arithmetic operations with a cost that grows slowly rather than linearly. It follows that the "right" measure to use depends on the particular case.

Let us now consider again the problem of finding the time and space costs for an algorithm solving a problem R on a given computing model. The cost will generally be a function of the particular instance. However, in some cases it is only a function of the instance size. Take the example concerning matrix multiplication (see Example 2.7). The number of instructions in the program shown in Example 2.7 is independent of the values of the matrices A and B, and is a function of n. The same considerations apply to a RAM program implementing the straight–line program. Under the unit cost hypothesis, these programs have a time cost dependent only on the value n.

When dealing with other kinds of problem, the cost is actually a function of the instance. As an example, searching an object in a set costs much less if the set has some kind of ordering. It is not very sensible however to define the cost as a function of the instances. Criteria must be found to be applied to the majority of cases. For instance, the average case and the worst case behaviour of the algorithm.

Let us first consider the average case. Obviously, each instance of a given problem has a certain probability of occurring. Each instance is associated one–to–one to a computation (we only consider deterministic algorithms) on any computing model. The average cost is computed by summing the cost of the computations on an input of size n, and then multiplying each cost by the occurrence probability of the corresponding instance. Let us formalize this concept with the following definition.

Definition 2.37
Let A be an algorithm solving a problem R on a computing model M, and let $C_n(M, A)$ denote the set of computations originated by A on the elements of R_n. For simplicity we only consider the case when $C_n(M, A)$ is countable. If p_j denotes the occurrence probability of computation $C_j, j = 1, 2, ...$, we define the following quantity as the *average time cost* (of A):

$$\overline{T}_A(n) = \sum_{C_j \in C_n(M, A)} T(C_j) p_j, \quad \sum_j p_j = 1$$

A similar definition applies to the average space cost. Note that the previous definition adopts the unit cost measure. A more correct definition should include the unit measure among its parameters. We have chosen not to include it for the sake of simplicity.

The average behaviour of an algorithm is key information. However, note that:

- it can be very difficult to determine the probability of the occurrence of each instance for a given problem;
- a "good" value of the average cost may hide some particular instances (sometimes very important ones) for which the algorithm is not well suited.

The worst case cost is used more frequently: it is mathematically easier to treat, and the information it provides, i.e. the worst that you can ever expect from a given algorithm, gives a more general indication (it can be applied to *each* instance of the problem).

Definition 2.38
Let A, M and $C_n(M, A)$ be defined as in Definition 2.37. We define the *worst case time cost* of algorithm A as the following quantity:

$$T_A(n) = \max_{C_j \in C_n(M, A)} T(C_j)$$

(Again, a similar definition applies to the worst case space cost.)

Example 2.39
Let $X = \{x_1, x_2, ..., x_n\}$ be a set of n distinct numbers. Let us consider the problem of determining the minimum value in X. An intuitive algorithm (later referred to as MIN) is the following one:

begin
    for $j \leftarrow 1$ **until** $n-1$ **do begin**   comment x_j is the *candidate* number;
    $k \leftarrow j + 1$;
    while $k \leq n$ **and** $x_j < x_k$ **do** $k \leftarrow k+1$;
    if $k = n + 1$ **then begin**   comment if x_j is the minimum value
                                     found, the program returns the value j;
        write j;
        stop
    end
    end
    write n;   comment if the **for** execution completes
                                        regularly, the minimum is x_n;
end

Let us calculate the worst case time cost of this algorithm. We only consider the total number of comparisons assuming that: (1) each comparison requires the same amount of time, independently of the compared numbers; (2) the time required by other operations (test and increment of the indices, etc) can be neglected with respect to the time required by comparisons. It is easy to realize that the worst case is when the minimum is x_n and the other numbers are disposed in increasing order, i.e. $x_i < x_{i+1}$, $i = 1, ..., n - 2$. In this case each number x_i, $i = 1, ..., n - 1$ is compared with all the following numbers. The maximum number of comparisons is then given by the following expression:

$$T(n)_{\text{MIN}} = \sum_{i=1}^{n-1}(n - i) = \sum_{i=1}^{n-1} i = \frac{n(n-1)}{2}$$

Example 2.40

The MIN algorithm of the previous example possibly performs many comparisons whose result is already determined from the previously obtained results. More precisely, if $x_i > x_j$, where x_i is the current candidate, it also holds that $x_k > x_j$, $k = i + 1, ..., j - 1$. In fact, if the candidate reaches the comparison with x_j, it has "passed" the tests with elements with an index less than j. This also means that elements $x_i, x_{i+1}, ..., x_{j-1}$, rather than only x_i, can be excluded from other comparisons. The following algorithm MIN2 takes these considerations into account.

```
begin
    j ← 1;
    while j < n do begin              comment x_j is the candidate number;
        k ← j + 1;
        while k ≤ n and x_j < x_k do k ← k + 1;
        if k = n +1 then begin        comment if x_j is the minimum value
                                              found, the program returns
                                              the value j;
            write j;
            stop
        end
        else begin                    comment otherwise the new candidate is x_k;
            j ← k
        end
    end
    write n ;                         comment if the while execution completes
                                              regularly, the minimum is x_n;
end
```

It is easy to verify that $T_{MIN2}(n) = \overline{T}_{MIN2}(n) = n - 1$. Note that an element is "discarded" if and only if it loses a comparison. Since the elements to be discarded are $n - 1$, the algorithm performs exactly $n - 1$ comparisons.

When there are no ambiguities, we denote the worst case time (resp. space) cost with $T(n)$ (resp. $S(n)$), i.e. disregarding the algorithm.

This book considers only worst case algorithm costs. This is because of the previous motivations and is in accordance with the great majority of the literature. However, it must be noted that in some cases the high cost of an algorithm is verified on "pathological" instances of the problem, or on very rare ones. An important example is the *simplex algorithm* solving linear programming problems. The worst case of the simplex algorithm has a cost which is an exponential function of the size (we will see in the next section that this is a very unfavourable situation); however, the average behaviour is still good, thus the algorithm is generally competitive.

2.4.2 Algorithm cost criteria: asymptotic behaviour

The time/space cost appropriateness of an algorithm is usually evaluated with respect to its asymptotic behaviour, i.e. when the size tends to infinity. In fact a "good" asymptotic behaviour guarantees that size increments correspond to a reasonable increase in the algorithm cost. Vice versa a bad asymptotic behaviour makes the algorithm applicable only to instances of limited size. From this point of view only algorithms with a cost which is at most a polynomial function of the size of the problem, i.e. $T(n) = n^{O(1)}$ and $S(n) = n^{O(1)}$, can be considered "good". Vice versa the algorithms with a cost which is an exponential function of the size of the instance to solve are considered to be prohibitive.

It must be clear that for some small size values, an exponential cost algorithm can be preferable to a polynomial cost algorithm. For example, even if:

$$\lim_{n \to \infty} \frac{n^{100}}{2^n} = \lim_{n \to \infty} \frac{100 \log n}{n} = 0$$

2^n is bigger than n^{100} only when $n \geq 997$. However, an algorithm with exponential cost is applicable only to instances of the problem with a limited size. Major technology developments will only lead to minor enlargements of the applicability set.

Example 2.41 (see also [AHU74])
Assume we have a machine capable of executing k instructions in the unit time, and let us adopt the unit measure cost. Let R be a given problem and let A be an algorithm solving R with time cost $T_A(n) = n^2$. A can manage in the unit time instances of R with a size n satisfying the relation $n^2 \leq k$, i.e.

$n \leq \sqrt{k}$. If we consider a second machine with twice the speed of the first, A can manage instances of size $n \leq \sqrt{2k} \approx 1.41\sqrt{k}$. With respect to the previous example we have obtained a 41% increase. Let us now consider an algorithm B solving R with cost $T_B(n) = 2^n$. When considering the first machine the maximum instance size manageable in the unit time is $n = \log k$, whereas with the second machine the maximum size is $n = \log 2k = \log k + 1$, i.e. the increment is a (small) quantity independent of k.

Note that the previous example provides a strong reason for pursuing the development of efficient algorithms. It in fact proves that algorithms, more than technology, define the boundary of treatable problems. The term technology refers to the aspects related to the physical construction of a machine (mainly space and CPU), and the great speed improvements in the succeeding computer "generations". The machine architecture is mainly the same as that of Von Neumann. Even if the speed increment is considerable, the previous example shows that the cost of certain problems can grow much more rapidly.

We have just seen two principles, that is:

(a) the asymptotic behaviour determines the range of applicability of an algorithm;
(b) the boundary between good algorithms and bad algorithms can be attempted by following the polynomial/exponential cost criteria.

The direct consequence of these two principles is that evaluation of algorithm analysis is generally performed by using magnitude orders, i.e. by using the notations O, Ω and Θ (see Appendix A). As an example, if $T(n) = kf(n) + g(n)$, $k = O(1)$, where $g(n) = O(f(n))$, then one writes $T(n) = O(f(n))$, to indicate that the cost growth order is dominated by the term $f(n)$. This kind of writing "hides" both additional additive contributions and multiplication factors of the dominant factor that can be important in practice.

Example 2.42
The cost of the algorithm MIN of Example 2.39 is $T_{MIN}(n) = O(n^2)$. In fact, the limit of the ratio $T_{MIN}(n)/n^2$ is finite (and equal to 1/2). The asymptotic notation then hides the multiplication factor 1/2 and the additive contribution $-n/2$. In this particular case neither is important for small values of n.

2.4.3 Problems: complexity bounds

Every algorithm A solving a problem R on some computing model M states that the intrinsic "difficulty" of R with respect to a solution on M is not greater than the time/space costs of A.[5] This fact leads to the following definition of problem complexity.

---

[5] In the following we shall often omit the reference to the computing model.

Definition 2.43
Let R be a problem and let $A(R)$ be the set of all algorithms solving R. The time complexity $C_T(R)$ (resp. space complexity $C_S(R)$) of R is the lower bound of the set of time (resp. space) costs of all $A(R)$ algorithms. Formally:

$$C_T(R) = \inf_{A \in A(R)} \{T_A(n)\} \quad \left(\text{resp. } C_S(R) = \inf_{A \in A(R)} \{S_A(n)\}\right)$$

The above definition encapsulates the intuitive idea of intrinsic difficulty of a problem; it cannot however be used in practice. Generally, we can say that the cost of an algorithm solving a problem represents an *upper bound* on its computational complexity.

Definition 2.44
Let A be an algorithm with time cost $T_A(n) = O(g(n))$ solving a problem R. The function g is an *upper bound on the time complexity of R*. A similar definition applies for the *upper bound on the space complexity of R*. The best *known* upper bound on time (or space) complexity is the minimum time (resp. space) cost among all the *known* algorithms solving a given problem. How can we know that an algorithm solves a given problem with the optimum cost?

Definition 2.45
Consider problem R. A function f such that $T_A(n) = \Omega(f(n))$ for each algorithm A solving R is called a *lower bound on the time complexity of R*. A similar definition applies to the *lower bound on the space complexity of R*.

In the following we will only refer to time complexity (the extension to space is obvious).

Let f be a lower bound on the complexity of R. Any other function ϕ such that $\phi(n) = O(f(n))$ is still a lower bound on the complexity of R. If $\phi(n) < f(n)$, ϕ is a worse bound than f (i.e. it gives less information on the actual difficulty of R). Among the lower bounds the concept of best *known* bound is still valid.

Let g be the best known upper bound on the complexity of a given problem, and let it be asymptotically comparable to the best known lower bound f, i.e. $g(n) = \Theta(f(n))$. Then, *and only then*, we can conclude: (1) an algorithm costing $T(n) = O(g(n))$ is *optimum*, and (2) the *computational complexity* of the problem is $\Theta(f(n))$. In this case the problem is definitely solved, at least from an asymptotic point of view.

There are not many problems where such a solution has been found. A complete knowledge of the computational complexity of a problem is actually very rare. Often the crucial problem is determining a good lower complexity bound. However, for some important problems the optimum cost algorithms have been found.

Example 2.46
Algorithm MIN2 of Example 2.40 is optimum. Given that MIN2 has a time cost $T_{MIN2}(n) = n - 1$, we must prove that $\Omega(n)$ comparisons are necessary to find the minimum in a set of n numbers. The proof is very simple. In fact, an element which is not the minimum must lose at least one comparison. Vice versa each comparison can determine the elimination of at most one element. It follows that $n - 1$ comparisons are necessary to discard the same number of elements.

Example 2.47
Any algorithm solving the sorting problem of n numbers (see Examples 2.8 and 2.25) requires $\Omega(n \log n)$ comparisons. For the sake of simplicity, let us suppose that the elements to be sorted, $x_1, x_2, ..., x_n$, are all distinct. We can use the decision tree as the computing model (see Figure 2.4). Each leaf of the tree has associated with it a possible solution to the problem. Hence there are at least $n!$ (i.e. the number of permutations of n distinct elements) leaves in the tree.[6] Moreover, the path connecting the root to a leaf represents the sequence of comparisons necessary when the solution is exactly the permutation associated with that leaf. We can determine a lower bound on the worst case complexity by analyzing the tree structure. Note that in a binary tree with k leaves the maximum path length is at least $\lceil \log k \rceil$. In our case the path length represents the number of performed comparisons. As the number of leaves is at least $n!$, the worst case of any sorting algorithm must perform at least $\lceil \log n! \rceil$ comparisons. By using the Stirling formula:

$$\lim_{n \to \infty} \frac{\sqrt{2\pi n}\, n^n e^{-n}}{n!} = 1$$

it follows that $\log n! = \Theta(n \log n)$ and hence the number of comparisons is $\Omega(n \log n)$.

Let us now summarize the criteria that will be used in this book in order to define algorithm cost and problem complexity:

- there is a reference computing model;
- the cost measure is defined;
- the worst case analysis is performed;
- the asymptotic case is studied.

---

[6]The leaves representing different permutations are exactly $n!$. However, it is possible that the algorithm performs "useless" comparisons (as in the case of the algorithm MIN in Example 2.35). Hence it is possible that different paths lead to the same solution, i.e. that different leaves represent the same solution.

2.4.4 Relationships among computing models

This section introduces the relationships between Turing machines and random access machines. Section 2.2.3 outlined how these models are equivalent with respect to computed functions. We now want to investigate the relationships with respect to used resources. The technique which permits us to establish relationships among models is the *simulation*.

Definition 2.48
The time complexity $T(n)$ (resp. space complexity $S(n)$) of a TM is the maximum time cost (resp. space) of all the computations on an input of length n. If, on an input of length n, at least one computation diverges, the value $T(n)$ (resp. $S(n)$) is undefined.

Time and space complexity for a RAM are defined in a similar way. Note that time complexity, both for TM and RAM, satisfies the relation $T(n) \geq n$. This is based on the assumption that during a computation the entire input is read (requiring at least n computation steps). On the other hand, having considered in a different way input and working space (see Section 2.2.2 on models) allows us to consider computations with a sublinear working space.

Example 2.49
The time complexity for the TM computing the sum of two integers of h and k bits (see Example 2.14) is $O(n)$, where $n = \max\{h, k\}$. Each computation is divided into three phases: (1) the first operand is copied from the input tape to the working tape; (2) the working tape head is positioned on the first bit of the first operand; (3) the addition is computed. It is easy to realize that the cost of the three phases is $O(h)$, $O(h)$ and $O(n)$ respectively. The time complexity is then $O(n)$. The space complexity is $h + 1$.

Proposition 2.50
A TM computation of time complexity $T(n) \geq n$ can be simulated by a RAM in $O(T(n))$.

Proof. The simulating RAM uses register r_0 to store the status of the simulated TM, and registers $r_1, ..., r_k$ to store the head positions on the tapes. The ith tape is represented by registers r_{b+kt+i}, $t = 0, 1, 2, ...$, where $b > k$. Registers r_{k+1} up to and including r_b are used as temporary working areas. The RAM can read the input and transfer it into the first n registers representing tape 1 in $O(n)$. Each TM step can be simulated in constant time (i.e. independent of n). Note that there is a constant number of control states and of possible k–tuples of symbols read from the tapes. Hence the RAM can determine the next status and the actions on the tapes in time $O(1)$. The heart of the simulation is the execution of control constructs as shown in Figure 2.11 (where s and t are states, x_i and y_i are symbols of the TM alphabet) describing a status transition. Translating the program shown in Figure 2.11 into a real program is just a matter of patience.

Vice versa the following proposition holds.

Proposition 2.51
A RAM computation with time complexity $T(n) \geq n$ can be simulated by a TM in time $O(T(n)^3)$.

Proof (sketch). The simulating TM uses five tapes. Tape 1 stores input, tape 5 stores output. The other tapes are working tapes. Tape 2 is used to store the content of RAM registers, tape 3 stores the content of the accumulator, and tape 4 is an additional working area. Recall (see the discussion of RAMs in Section 2.2.2) that in a computation of time cost $T(n)$, only $O(T(n))$ RAM registers can contain a value other than zero. These registers are represented on TM tape 2 as sequences of pairs <address, value>. This is done by using special characters to separate pairs, and to separate the address item from the value item. Let us now suppose that the length in bits of each of the n input data is at most n. This means that the length of tape 1 is at most n^2. As only addition and subtraction are present in the RAM instruction set, a computation with cost $T(n)$ can generate numbers with a value not bigger than $2^{n+T(n)}$, which can be represented using $O(T(n))$ bits. It follows that the portion of tape 2 containing symbols other than blank is $O(T(n)^2)$. This is also the cost of simulating a single RAM instruction. As an example, let us consider the instruction "ADD ri". The TM must first look for a pair on tape 2 having i as its first element. Then it has to copy the second element of this pair (i.e. the first operand) to tape 4. The second operand (i.e. the accumulator) is on tape 3. The sum execution has a cost proportional to the maximum of the two operand lengths (see Example 2.49), and hence $O(T(n))$. When the sum is computed the result is copied to tape 3, replacing the old value. The cost of

```
...
if    r₀ = s and @r₁ = x₁ and ... and @rₖ = xₖ
      then begin
           r₀ ← t ;                    comment t is the next state;
           @r₁ ← y₁ ;                  comment write on the first tape;
           ...
           @rₖ ← yₖ ;                  comment write on the kth tape;
           r₁ ← r₁ + k ;               comment (right) shift the first head;
           ...
           rₖ ← rₖ - k ;               comment (left) shift of the kth head;
      end;
...
```

Figure 2.11 Simulation of a TM computation step.

the simulation is then determined by the search on tape 2, and hence it is $O(T(n)^2)$. As the number of operations to simulate is $T(n)$, the total simulation time is $O(T(n)^3)$.

Definition 2.52
Two functions f and g defined on N are *polynomially related* if there exists a positive constant k such that, for each n, $f(n) = O(g(n)^k)$.

Definition 2.53
Two computing models M_1 and M_2 are *polynomially related* if each computation $c \in C(M_1)$ can be simulated by a computation $d \in C(M_2)$, and if the functions $T(c)$ and $T(d)$ are polynomially related.

It follows from Propositions 2.50 and 2.51 that "Turing machines and random access machines are polynomially related computation models".

Sequential computation thesis

We have already summarized in Section 2.4.2 that a possible criterion for algorithm classification based on complexity is the following: "efficient algorithms have a cost which is a polynomial function of the dimension". We now want to discuss further this criterion, called *polynomial efficiency*, keeping in mind that certain costs, e.g. n^{1000}, make algorithms unusable in practice.

The main reason is that a theory of algorithms based upon polynomial efficiency is independent of the machine. Most sequential computing models are polynomially related, hence it makes sense to deal with *algorithm efficiency* in general, i.e. without referring to a particular model. The fact that each computation costing $T(n)$ in a modern computer can be simulated by a TM in $T(n)^{O(1)}$ could cause some doubts. The result that TMs and RAMs (the latter are surely more similar to real computers) are polynomially related could not represent a convincing element. One could argue that the RAM instruction set does not include multiplication and division, i.e. two operations present in modern computers. We partially gave an answer to this objection in Example 2.5, showing that multiplication can be performed in polynomial time by means of the sole addition. If this is not considered to be enough, we can see what happens, depending on the cost measure, when multiplication is part of the instruction set.

(a) Unit cost. A RAM with multiplication can compute, in a computation with length $T(n)$, numbers with value $O(2^{2^n})$ requiring $O(2^n)$ bits to be represented. A TM simulation of any arithmetic instruction has at least a cost linear with the length of the operand representation (see Example 2.49). It follows that no polynomial relationship can exist between the cost of a computation on *this* RAM and the cost of a similar computation on a TM. However, in this case the RAM and not the TM is an unrealistic model. In fact, even the most powerful computer does not have memory locations that contain a number with 2^n bits (recall that we always consider the asymptotic case). If one has to split the representation, the instruction cost will depend on the number of locations where the operands are split. Let us now suppose that each location can contain $O(n^k)$ bits, $k = O(1)$. It follows that the number of locations necessary to store an operand with 2^n bits is $\Omega(2^n/n^k)$, which is still a function growing exponentially with n.

(b) Logarithmic cost. In this case a RAM with multiplication is again polynomially related with a TM (see Proposition 2.54).

Proposition 2.54
A computation of a RAM including multiplication in its instruction set, with time complexity $T(n) \geq n$ with respect to logarithmic cost, can be simulated by a TM with time complexity $O(T(n)^k)$, $k = O(1)$.

Proof. See [AHU74].

The sequential computation thesis [GL83] expresses the idea that "all future computing models will be polynomially related to Turing machines". Obviously, this is an extension of the Church–Turing thesis.

2.5 Complexity classes, reductions, completeness

We have seen that the computational complexity theory based on the concept of polynomial efficiency has several advantages. However modern research does provide a classification of problems (and algorithms) more accurate than the mere distinction between polynomial time and exponential time. In the following we will describe concepts and results which clearly show a more articulated situation.

This section considers only decision problems (see Section 2.3) and refers to the TM as the underlying computing model. A TM for a decision problem does not have an output tape. Since the result of any computation belongs to a set of two values ({YES, NOT}), it is sufficient for the machine to have two final distinct states, called *accept* and *reject*. We use these terms because a decision problem $R \subseteq D \times \{YES, NOT\}$ can be seen as the problem of recognizing the language R_Σ composed of all the strings on the alphabet Σ coding elements belonging to D (i.e. instances of R) whose solution is YES. We can use this analogy between decision problems and language recognition, saying that M *accepts language R* (for simplicity we omit the alphabet notation) instead of saying that M solves the decision problem R.

An additional simplification (used in this section as well as in the next one) is to consider TMs with only one working tape. This restriction is not crucial.

In fact, the following result holds.

Proposition 2.55 (theorem on tape number reduction)
The computations of a TM M of time complexity $T(n)$, with $k \geq 2$ working tapes, can be simulated by a TM M' of time complexity $O(T^2(n))$, using only one working tape.

Proof. See [AHU74].

It is easy to realize that a constant reduction in the number of tapes does not involve any increase (in the order of magnitude) of space requirements.

A key concept in computational complexity is that of *complexity class*, i.e. the set of all problems which can be solved on a given computing model with precise limitations on the use of one or more computing resources. Let S and T be positive non-decreasing functions defined on the natural numbers. We define:

$\text{SPACE}_{\text{TM}}(S(n)) = \{R: \exists$ a TM with space complexity $O(S(n))$ accepting $R\}$,
$\text{TIME}_{\text{TM}}(T(n)) = \{R: \exists$ a TM with time complexity $O(T(n))$ accepting $R\}$.

Let us also consider classes of problems with simultaneous resource limitations:

$\text{TIME-SPACE}_{\text{TM}}(T(n), S(n)) = \{R: \exists$ a TM with time complexity $O(T(n))$ and space complexity $O(S(n))$ accepting $R\}$.

Consider finally complexity classes related to nondeterministic computations:

$\text{NSPACE}_{\text{TM}}(S(n)) = \{R: \exists$ a non-deterministic TM with space complexity $O(S(n))$ accepting $R\}$,
$\text{NTIME}_{\text{TM}}(T(n)) = \{R: \exists$ a non-deterministic TM with time complexity $O(T(n))$ accepting $R\}$.

The following question arises: which are the meaningful complexity classes, i.e. those containing interesting problems? Concerning space complexity it is obvious that, for each space limitation S, $\text{SPACE}_{\text{TM}}(S(n)) \subseteq \text{NSPACE}_{\text{TM}}(S(n))$. Moreover, if $S'(n) \leq S(n)$, then $\text{SPACE}_{\text{TM}}(S'(n)) \subseteq \text{SPACE}_{\text{TM}}(S(n))$. What happens if $S'(n) < S(n)$? We need the following definition.

Definition 2.56
A function f defined on N is said to be *space constructible* if there exists a TM with space complexity f that, given an input with length n, writes a special symbol in the $f(n)$th cell of its working tape.

As we have seen in the previous section, many complexity results are obtained by means of simulations. While executing a simulation it is often necessary to have a preset working space, usually depending on the input size. The need for a working space $f(n)$ can be satisfied when f is a space constructible function. The most common functions, such as $\log n$, n^k and 2^n, are space constructible.

Proposition 2.57 (space hierarchy theorem [HLS65, HU79])
If S is space constructible and:

$$\lim_{n \to \infty} \frac{S'(n)}{S(n)} = 0$$

then $\text{SPACE}_{\text{TM}}(S'(n)) \subset \text{SPACE}_{\text{TM}}(S(n))$, i.e. there exists at least a problem in $\text{SPACE}_{\text{TM}}(S(n))$ not belonging to $\text{SPACE}_{\text{TM}}(S'(n))$.

Proof (sketch). Let M' be a TM with space complexity $S'(n)$. Let us consider a TM M with complexity $O(S(n))$ working in the following way. M gets an input string $w = f\#x$, where f is a coding of the transition function of M' and w is the input for M' ($|w| = n$). M reserves a working tape portion composed of $S(n)$ cells (this is possible because S is space constructible). It then simulates M' on input n using the reserved tape portion.[7] If M' accepts x then M rejects x; vice versa, if M' rejects, or if its computation requires more space than reserved,[8] then M accepts.

Because of the limiting condition, for a sufficiently large value of n, the simulation of *all* the TM with space complexity $S'(n)$ can be performed with the reserved space. As a consequence when an input has such a length, the response of M differs from the response of any TM with space complexity $S'(n)$.

A similar result applies also to time (but the proof is more complicated).

Proposition 2.57 seems to suggest that significant complexity classes for space (or time) can be obtained by considering sequences of functions $\{S_i\}_{i \in N}$ such that $S_i < S_{i+1}$, for each i. The interest is actually concentrated upon a limited number of function classes.

As far as space is concerned, very interesting classes are composed of

---

[7] Note that M is partly composed of the universal Turing machine (see par. 2.2.3).
[8] The fact that definitely $S'(n) < S(n)$ does not exclude that, with "sufficiently small" values of n, is $S(n) \leq S'(n)$.

problems which can be solved in (poly)logarithmic space, i.e. POLYLOGSPACE = SPACE$_{TM}$((log$n$)$^{O(1)}$). In addition to its interest for sequential computations, POLYLOGSPACE is important because it contains problems that can be solved very fast in parallel (see Chapter 3). Another important class is composed of the problems which can be solved in deterministic polynomial space, i.e. PSPACE = SPACE$_{TM}$($n^{O(1)}$)).

Regarding time we have already noticed that a precise characterization of the problems that can be solved (at most) in polynomial time is essential. These problems are treatable in practice. We then define P = TIME$_{TM}$($n^{O(1)}$)). Also the class NP = NTIME$_{TM}$($n^{O(1)}$)) is very important. NP is the class of problems that can be solved by a non–deterministic TM in polynomial time. Recall that a non–deterministic machine, when facing a problem for which the exhaustive search is the best known strategy, (1) chooses a candidate configuration and (2) verifies that it is really a solution. NP contains searching problems that, once the lucky candidate configuration has been chosen, have a verification stage that can be performed in polynomial time (with respect to problem size).

Example 2.58 (DCLIQUE(k))
Let us consider the problem of deciding whether a k–clique exists in a graph with n nodes (see Example 2.29). The number of possible k–cliques in the graph is exactly the number of subsets of k elements of a set with cardinality n. This value is $\binom{n}{k}$, a quantity which is generally not dominated by any polynomial function in n. The problem resides in NP. In fact, a non–deterministic TM can select in constant time a configuration with k nodes in the graph, and verify in polynomial time if it is really a k–clique. Note that verifying that a set of k nodes is completely connected can be performed in time O(k^2).

As shown in the previous example, non–determinism represents a useful abstraction in order to characterize classes of problems. In theory NP problems may still have deterministic polynomial algorithms. However, we shall see in the following that this possibility is very unlikely at least for a certain subset of NP.

The classes introduced above have the following relationships:

$$P \subseteq NP \subseteq \text{PSPACE} \tag{3}$$

The relationship $NP \subseteq$ PSPACE is not trivial. It depends on a result obtained by Savitch on the space necessary to simulate non–deterministic computations [Sa70]. Let us first consider the following example.

Example 2.59 (GAP)
The accessibility problem on a directed graph $G = (V, E)$ (see Example 2.30)

can be solved by a non–deterministic TM in space complexity O(log n), and by a deterministic TM in space complexity O($\log^2 n$), i.e. GAP ∈ NSPACE$_{TM}$(log n) and GAP ∈ SPACE$_{TM}$($\log^2 n$). Consider the following non–deterministic algorithm, where s and t are two nodes whose connectivity has to be verified.

```
begin
    X ← s;
    while X ≠ t do begin        comment choose in a non–deterministic way a node
                                         Y;
        if (X, Y) ∈ E then
            X ← Y
        else begin
            write NO;            comment there is no path between s and t;
            stop end
    end;
    write YES;                   comment there is a path between s and t;
end
```

If the nodes in V are represented, as usual, with the numbers from 1 to n, the space necessary to implement the above algorithm with a non–deterministic TM is O(log n). In fact, the only things that have to be written on the working tape at each step of the computation are the contents of variable X and the node (represented by V) chosen non–deterministically as the successor of X.

The previous algorithm can be simulated by the following recursive function (see [Pa87]).

```
procedure PATH (I,J,K);  comment return true if there is a path with length at
                                  most K, connecting node I to node J;
                                  otherwise return false.
begin
    if I = J then return true
    else if K = 1 return (I, J) ∈ E
    else return ∃L: PATH(I, L, ⌊k/2⌋) and PATH(L, J, ⌈k/2⌉)
end
```

The expression:

∃L: PATH(I, L, ⌊k/2⌋) and PATH(L, J, ⌈k/2⌉)

can be verified by trying all the possible vertices L of the graph. Using the procedure PATH, the accessibility problem can be solved in the following way:

if PATH$(s, t, n-1)$ **then** write YES
else write NO

Note that the recursion depth is at most $\lceil \log n \rceil$, and at each step a constant number of variables is used. Their content requires $O(\log n)$ bits to be represented (it is a number between 1 and n). It follows that the total space requirements for the procedure PATH is $O(\log^2 n)$.

Proposition 2.60 [Sa70]
If $S(n)$ is space constructible, then:

$$NSPACE_{TM}(S(n)) \subseteq SPACE_{TM}(S^2(n))$$

Proof. A (non–deterministic) TM using $S(n)$ cells of working tape has $N = qnc^{S(n)}S(n)$ configurations, where q is the number of states and c is the number of alphabet symbols (both are constant values). The previous value is obtained by considering that a configuration is determined by the control status, the head position on the input tape, the contents of the working tape, and its head position. Hence each configuration can have a number between 1 and N associated with it. We can force the TM to have only one acceptance configuration. In order to do this, the machine will erase its working tape and position the head on the first cell before entering the acceptance state. Configurations can be considered as nodes in a digraph, called the *computation graph*. Two nodes i and j are linked by an arc if and only if the machine can go from the configuration corresponding to node i to the one corresponding to node j in a single step. The problem of determining if a TM with space complexity $S(n)$ accepts an input with length n is equivalent to the problem of determining if, in the computation graph, the node corresponding to the final configuration is reachable (accessible) from the node corresponding to the initial configuration. Since the graph has N nodes, it follows from Example 2.59 that the problem can be solved in space $O(\log^2 N) = O(S^2(n))$.

From the above we have NPSPACE = PSPACE, where NPSPACE = $NSPACE_{TM}(n^{O(1)})$. Moreover, as a TM (either deterministic or non–deterministic) with time complexity $T(n)$ cannot use more than $T(n)$ cells, it is $NP \subseteq NPSPACE$. By combining these two facts the rightmost part of (3) is obtained.

Let us now introduce the concept of transitive closure of a graph. This concept is strictly linked to the accessibility problem. The concept will be used in Chapter 3.

Definition 2.61
Let G be a digraph. The graph G^* having the same set of vertices as G, and having the arc (i, j) if and only if there exists a path in G linking i to j, is called the *transitive closure* of G.

(a)

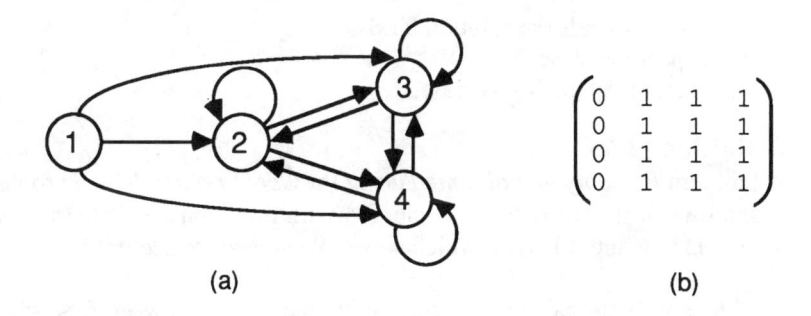

$$\begin{pmatrix} 0 & 1 & 1 & 1 \\ 0 & 1 & 1 & 1 \\ 0 & 1 & 1 & 1 \\ 0 & 1 & 1 & 1 \end{pmatrix}$$

(b)

Figure 2.12 Transitive closure of the graph in Figure 2.9.

The transitive closure of the graph in Figure 2.9 and its representation by means of the adjacency matrix are described in Figure 2.12.

It clearly follows from Definition 2.61 that the GAP can be solved, for each pair s and t, by computing the transitive closure of the graph. Vice versa, by solving the GAP for each possible pair of nodes, the closure of the graph is computed. The transitive closure of a graph G can also be determined by computing the $(n-1)$th[9] power of its adjacency matrix A. Note that the non–vanishing elements in the matrix $A^2 = A \times A$ describe paths in G at most of length 2. Hence if the element of A^2 with index (i, j) is 1, then there exists in G a path of length at most 2 linking i with j. Similarly, the matrix A^k, with $k > 2$, describes paths of length at most k in G. As the maximum length of a path (without loops) in a graph with n nodes is $n-1$, the matrix A^{n-1} describes the transitive closure G^* of G. We will see in chapter 3 how the computation of the transitive closure of a graph plays a fundamental role in simulating sequential computations on parallel computing models.

Up to now it is not known whether the inclusions in (3) are proper, even if this is the current line of thought. The existence of problems that complete for *NP* (called NP–complete) makes the equality $P = NP$ most unlikely. In order to understand the completeness concept it is necessary to have the reducibility concept clear in mind.

Definition 2.62
Let us consider a (decision) problem R. An *oracle TM* (with oracle on R) is a TM with an additional tape N_Q, and with three additional states, named q_{ASK}, q_Y and q_N. The behaviour of the machine is similar to that of a traditional TM. However, if the computation enters status q_{ASK}, the next status will be:

1. q_Y, if the string on tape N_Q codifies an instance of R to which

---
[9]The adjacency matrix is seen as a Boolean matrix, hence the operations are logic operation (i.e. $1 + 1 = 1$).

corresponds the solution YES;
2. q_N, otherwise.

In both cases the tape N_Q is erased.

Definition 2.63
A problem R is *polynomial–time Turing* (or *Cook*) *reducible* to a problem R' if and only if there exists an oracle TM with time complexity $T(n) = n^{O(1)}$, where n is the input length, which solves R using an oracle on R'.

If R is polynomial–time Turing reducible to R', we write $R \leq_p R'$. Note that the relation $\leq_p$ is transitive, as the composition of polynomial functions is still a polynomial function. For the same reason, if $R \leq_p R'$ and $R' \in P$, then $R \in P$.

Definition 2.64
A problem R is polynomial–time Turing complete for NP (or simply NP–complete[10]) if it is in *NP* and if, for any other problem Q in *NP*, it holds $Q \leq_p R$.

If a polynomial–time deterministic algorithm is discovered for any NP–complete problem, we shall automatically have polynomial–time deterministic algorithms for any problem in *NP*, proving that $P = NP$. This is the most important proof in line with the assumption that P is strictly contained in *NP* (see Figure 2.13, where NPC indicates the class of NP–complete problems).

The set of NP–complete problems is a large one, containing very interesting computational problems. As an example, consider DCLIQUE(k). The first problem for which *NP* completeness has been proved is that known by the term *satisfiability*, or SAT more simply: "Given a Boolean formula in normal form,[11] is there a set of variable assignments such that the formula value is *true*?"

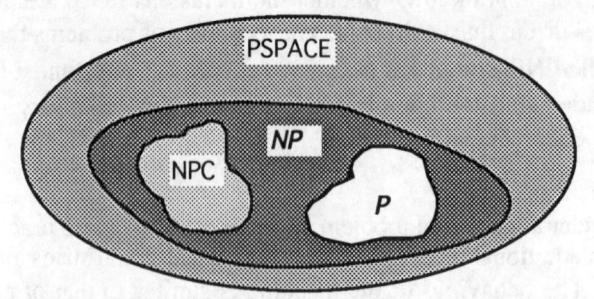

Figure 2.13 Structural hypothesis for some of the most important complexity classes under PSPACE.

---

[10] Actually the definition of a NP-complete problem involves a more restrictive notion of reduction, called polynomial transformation.
[11] A Boolean formula is in normal form a conjunctive (logic) product of (logic) sums.

NP completeness theory applies to decision problems just for convenience. However, a decision problem *DR* often corresponds to a general (constructive) problem *R*. In other words, *DR* is the decision version of *R*. As an example, before defining DCLIQUE(*k*) we defined CLIQUE(*k*) as the problem of determining a subset of *k* completely connected nodes (see Section 2.3). Similarly, there is both a decision and a non–decision version of HAMILTON. The decision version is another example of an NP–complete problem. Non–decision versions are often defined as *optimization* problems. As an example, consider the problem of "determining the maximum integer *k* such that the graph *G* has a clique with size *k*". It is an optimization problem, indicated with CLIQUE. In general, solving a problem *R* can be useful in solving the corresponding decision problem *DR* (e.g. DCLIQUE(*k*)). In this case if *DR* is NP–complete, then problem *R* is called NP–hard.

We have already noticed that the existence of complete problems for *NP* represents the most important proof in favour of the hypothesis $P \neq NP$. The same technique can be used to support another hypothesis, that is, (POLYLOGSPACE $\cap$ *P*) $\neq$ *P*, which is especially important with respect to its implications with parallel complexity. Let us now consider the following definitions.

Definition 2.65
Let *R* and *R'* be decision problems. We say that *R* is *logarithmic–space reducible* to *R'* if there is a function *f*, computable by a deterministic TM with space complexity $S(n) = \log n$, where *n* is the input length, such that *x* codes an instance of *R* with solution YES if and only if *f(x)* codes an instance of *R'* with solution YES. In such a case we write $R \leq_{\log} R'$.

The previous definition assumes that there is a language Σ on which the instances of *R* and *R'* are coded, hence $f: \Sigma^* \to \Sigma^*$. The relation $\leq_{\log}$ satisfies the following properties (the proof in not as easy as in the case of $\leq_p$):

1. if $R \leq_{\log} R'$ and $R' \leq_{\log} R''$, then $R \leq_{\log} R''$ (i.e. $\leq_{\log}$ is transitive);
2. if $R \leq_{\log} R'$ and $R' \in \text{SPACE}_{\text{TM}}(\log n)$, then $R \in \text{SPACE}_{\text{TM}}(\log n)$.

Hence we can formulate a definition, similar to 2.64, characterizing the "difficult" (with respect to space complexity) problems in *P*.

Definition 2.66
A problem *R* is complete for *P* (or simply P–complete) if for any other problem *Q* in *P* it holds $Q \leq_{\log} R$.

If there is a P–complete problem solvable in logarithmic space, then $P = \text{SPACE}_{\text{TM}}(\log n)$. The existence of complete problems for *P* represents the most important proof in favour of the fact that not all the polynomial time solvable problems are (simultaneously) solvable in logarithmic space. The first P–complete problem found is known as *Path System Accessibility* [Co74]. Given a set X, a relation $A \subseteq X \times X \times X$, and a subset S of X, an element $x \in X$ is defined *accessible* if $x \in S$ or if there exist two accessible nodes y and z such that $(x, y, z) \in A$. This problem can be formulated in the following way: "Given any subset T of X, is there an element of T which is accessible?"

Let us conclude noting that in [Jo75] a more general result regarding P–complete problems has been proven.

Proposition 2.67

If R is P–complete and if $R \in \text{SPACE}_{TM}((\log n)^k)$ for some k, then $P \subseteq \text{SPACE}_{TM}((\log n)^k)$.

Proof. See [Jo75].

Proposition 2.65 is the basis of the hypothesis $\text{TIME-SPACE}_{TM}(n^{O(1)}, (\log n)^{O(1)}) \neq P$ (i.e. $P \cap \text{POLYLOGSPACE} \neq P$) previously defined. We will see in the next chapter that the P–complete problems are the most difficult to parallelize.

Bibliographic notes

Even if some questions, such as $P =? NP$, still remain fundamental open problems (the solution of which could provide a totally new momentum), the complexity of sequential computation is a well established field. Several textbooks exist, some not quite recent, discussing this subject in a rather complete way (excluding the more specialized aspects) with respect to the present state of knowledge.

For clarity and breadth, the book by Aho, Hopcroft and Ullman [AHU74] is still one of the best introductory textbooks to algorithms and complexity. It analyzes, besides sequential computation models and basic techniques to synthesize efficient algorithms, algorithms for important computational problems arising in various applied arithmetic, numerical and combinatorial areas. The book also contains an excellent introduction to the theory of NP completeness.

For numerical and algebraic problems, the basic reference is the book by Borodin and Munro [BM75].

The complexity of important combinatorial problems, optimization problems in particular, is analyzed in detail in the book by Papadimitriou and Steiglitz [PS82].

The theory of NP completeness has its roots in the works by Cook and Karp, [Co71] and [Ka72]. A wide ranging discussion is also contained in the classic textbook by Garey and Johnson [GJ79], which provides an ample list of NP–complete problems. A list of P–complete problems, on the other hand, is given, together with the proofs of completeness, in the work by Jones and Laaser [JL77]. The first P–complete problem has been presented by Cook [Co74].

An introduction to automata theory and to the study of space and time hierarchies is contained in the classic textbook by Hopcroft and Ullman [HU79].

Problems

2.1 Design a deterministic Turing machine that recognizes the language $P_2 = \{ww^R \ w \in \{0,1\}^*\}$, where w^R equals w in reverse order.

2.2 Design a deterministic Turing machine that computes the product of two signed integers. Determine time and space complexity of your machine.

2.3 Give a RAM program to compute the following integer functions:
 (a) $f(n) = \lceil \log n \rceil$ ($f(n) = \lfloor \log n \rfloor$)
 (b) $f(n) = n!$
 (c) $f(n, m) = n$ div m ($f(n, m) = n$ mod m).

2.4 Prove Proposition 2.54.

2.5 Prove that the following functions are space constructible:
 (a) $f(n) = \lceil \log n \rceil$ ($f(n) = \lfloor \log n \rfloor$)
 (b) $f(n) = n^2$
 (c) $f(n) = 2^n$.

2.6 Given a language (or, equivalently, a decision problem) L over the alphabet $\{0,1\}$, let co–L denote the complement of L:

co–$L = \{x \in \{0,1\}^* : x \notin L\}$

Now, for a resource–bounded complexity class R, define:

co–$R = \{L : $ co–$L \in R\}$

For instance

co–TIME$(n) = \{L : \exists$ a DTM M with time complexity $O(n)$ that accepts co–$L\}$

Prove that the complexity class P is closed under complementation, i.e. that $P = $ co–P. Thus any DTM that accepts a language $L \in P$ is said to *decide P*.

2.7 Try to apply the same argument you (probably) used to solve Exercise 2.6 to prove that $NP = \text{co-}NP$. Explain the reason of your (very likely) failure. (Hint. Consider the one-sided nature of NP. The question $NP = \text{co-}NP$ is still unresolved, but it is strongly believed that $NP \neq \text{co-}NP$.)

2.8 Prove that $P \subseteq NP \cap \text{co-}NP$. (See Exercise 2.6 for the definition of the complement of a complexity class.)

2.9 Prove that log–space reducibility (Definition 2.65) is transitive. (Hint. You cannot simply compute $f(x)$ and apply the decision procedure for R' to $f(x)$, because $|f(x)|$ can be much longer than $|\log x|$. The key idea is to write a subroutine that, on input x and an integer i, produces the ith bit of $f(x)$.)

Chapter 3
Parallel Computation Models

As in the sequential case the notion of parallel computation can be formalized by defining some models. This chapter will illustrate the main models of parallel computation. These models are then used to introduce ideas and results that will form the basis of a *theory of parallel computation*. The chapter begins by defining the fundamental requirements of parallel computation, and by enunciating the *thesis of parallel computation* (Section 3.1). In Section 3.2, beginning with a detailed analysis of the relations between instructions in a sequential program, we will give a definition of parallel computation and introduce a first informal model of parallel computation. In the following sections we will analyze in detail a few formal models, starting with *Boolean circuits* up to *models of distributed systems*. We will study the relations, similar to those between models of sequential computation, connecting computations performed according to various models. Furthermore, we define important complexity classes for parallel computation, that is, classes of problems that can be solved by using a given amount of computational resources. In particular, we turn our attention to problems for which the theory predicts the existence of parallel algorithms exponentially faster than the best sequential ones.

3.1 Thesis of parallel computation

It is well known that space and time are the two most important resources characterizing the complexity of sequential computation (see Chapter 2). *Time* is still important in parallel computations: the main reason for studying parallel computation is the attempt to "speed up" the solution of computational problems. From an intuitive point of view one can think that with more processing units available the time needed to solve a given problem will be less. This fact points at processing units as the second main requirement for parallel computation. Referring to a model we will speak, rather than of some number of processing units, of an available quantity of *hardware*. The latter is a more general unit of measurement (which can, in some cases, coincide with the number of processing units).

The idea that time and hardware are the fundamental measures for parallel computation is backed by several results linking the sequential resources of space and time respectively with time and hardware in the parallel case. The correspondence found between sequential space and time in any "reasonable" model of parallel computation has given rise to the so-called *thesis of parallel computation*. Using the fact that all models of sequential computation are polynomially related to each other (Section 2.4), we can express the thesis of parallel computation as follows: any Turing machine computation with space cost $S(n)$ can be simulated in a reasonable model of parallel computation in time $T(n) = O(S(n)^c)$; conversely, any computation in a reasonable model of parallel computation with a time cost $T'(n)$ can be simulated by a Turing machine in space $S'(n) = O(T'(n)^d)$, where c and d are model-dependent constants.

By accepting the thesis of parallel computation there immediately follows a series of results which descend from known facts of sequential complexity.

Example 3.1
Consider a problem $R \in \text{TIME-SPACE}_{TM}(n^k, \log^h n)$, with k and h suitable constants. There are many problems with these characteristics, such as context-free language recognition. The thesis of parallel computation ensures that there exists a parallel algorithm which solves R in time $\log^{k'} n$, for a given k'. Comparing the sequential time $O(n^k)$ with the parallel time $\log^{k'} n$ one can classify R as an efficiently parallelizable problem (from the time point of view).

Example 3.2
Consider the class of P-complete problems (see Section 2.5). It is unlikely that a P-complete problem Q can be efficiently parallelized (in the sense of the previous example). If there existed a parallel algorithm solving Q in $O(\log^k n)$ time, then, according to the thesis of parallel computation, there would exist a sequential algorithm with space cost $O((\log^{k'} n))$ solving the same problem. On the other hand, due to the completeness of Q, this would mean that all problems in P can be solved in polylogarithmic space. That is, one would have $P \cap \text{SPACE}_{TM}((\log n)^{O(1)}) = P$, while the present assumption is that $P \cap \text{SPACE}_{TM}((\log n)^{O(1)}) \subset P$.

3.2 Parallel computations and algorithmic models

Most high-level programming languages (above all the oldest ones), although independent of a specific machine, have been designed keeping in mind the structure of a conventional computer (see Section 2.1). These languages are called imperative, since the programs they define are basically sequences of commands to the processor. It is the structure of the conventional computer itself (and its abstract models such as RAM) that forces a sequential view of computations.

From a logic point of view some program lines, even if they are physically in a sequence, can be thought of as independent. Consider once again the program that multiplies matrices of order n in Example 2.7 (shown in Figure 3.1(a)). A straight-line program corresponding to the case $n = 2$ is shown in Figure 3.1(b).

```
for i ← 1 until n do
    for j ← 1 until n do begin
        c(i,j) ← 0;
        for k ← 1 until n do
            c(i,j) ← c(i,j) + a(i,k) * b(k,j)
    end;
```

(a) Pidgin ALGOL program

```
1   c(1,1) ← 0;
2   c(1,1) ← c(1,1) + a(1,1) * b(1,1);
3   c(1,1) ← c(1,1) + a(1,2) * b(2,1);
4   c(1,2) ← 0;
5   c(1,2) ← c(1,2) + a(1,1) * b(1,2);
6   c(1,2) ← c(1,2) + a(1,2) * b(2,2);
7   c(2,1) ← 0;
8   c(2,1) ← c(2,1) + a(2,1) * b(1,1);
9   c(2,1) ← c(2,1) + a(2,2) * b(2,1);
10  c(2,2) ← 0;
11  c(2,2) ← c(2,2) + a(2,1) * b(1,2);
12  c(2,2) ← c(2,2) + a(2,2) * b(2,2);
```

(b) Straight-line program corresponding to the case $n = 2$

Figure 3.1 Order n matrix multiplication.

From an inspection of the code in Figure 3.1(b), it is clear that some instructions are independent of others, and therefore that the suggested sequence is a choice rather than a necessity. In particular, the program in Figure 3.1(b) can be divided into four different parts, each one of which calculates n elements of the product matrix. It is apparent that one could, for instance, calculate first the element $c(2,2)$, and then the element $c(1,1)$. Even while calculating one of these elements, the order of the instructions could be changed, executing for instance instruction 3 before instruction 2. There is however a difference between the two cases just discussed. The computation

62 Parallel computation models

of elements c(1,1) and c(2,2), as well as of c(1,2) and c(2,1), can be done in a parallel fashion (assuming the availability of suitable computing resources). On the other hand, in order to store correctly the final result in the variable c(1,1), some order, whatever it might be, has to be kept between instructions 2 and 3. Let us clarify these concepts with the help of a graphic formalism.

To denote that instruction i precedes instruction j we will use the following notation:

To indicate that instructions $i_1, ..., i_k$ precede instruction i, and that, conversely, instructions $j_1, ..., j_h$ are preceded by j, we write respectively:

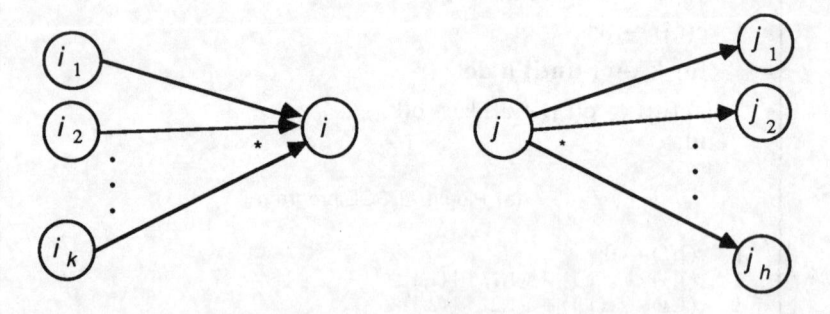

Using this formalism we can specify, in graphical form, a possible order in which program instructions are executed. Such a notation, shown in Figure 3.2, is called a *partial ordering graph.*.

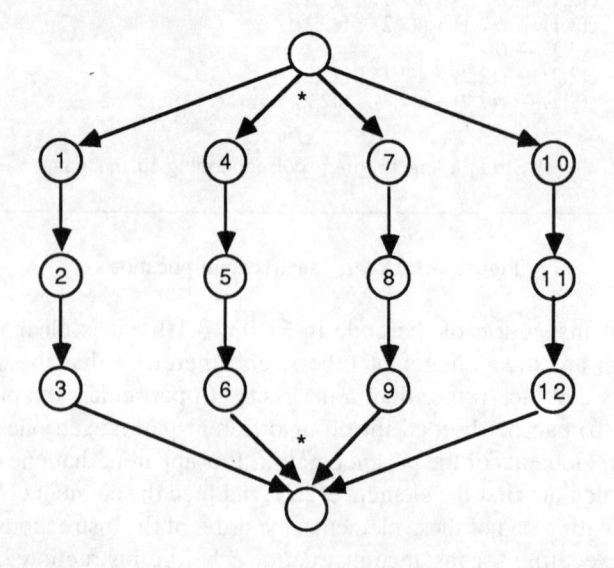

Figure 3.2 Partial ordering graph for the program of Figure 3.1(b).

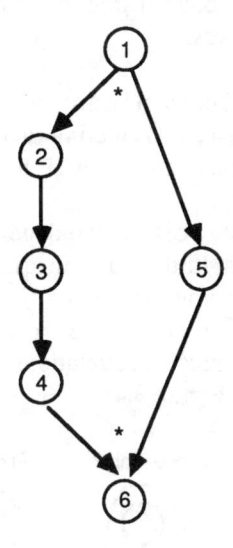

Figure 3.3 A partial ordering graph.

Two instructions not connected by a path on the graph are independent, and can be executed in parallel. It is clear that whenever a program flow is parallelized, resulting flows must be synchronized. This is necessary to ensure a correct computation. Consider for instance the partial ordering graph shown in Figure 3.3.

Assuming that each instruction requires the same time, the path (1,2)(2,3)(3,4)(4,6) will be "longer" than the path (1,5)(5,6). If the instructions along the two paths are actually executed in parallel, some synchronization mechanism between the two program flows must be provided to ensure that instruction 6 is not executed before instructions 4 and 5 have been completed. A simple solution would be to make all paths of equal length by inserting null instructions, for instance:

sync ← *sync*

where *sync* is any variable. To clarify the purpose of these instructions, we will use in such cases a "fake" instruction denoted by the code *nop*, standing for "null operation".

A partial ordering graph of the instructions of a program defines a possible *parallel computation*. It is obvious that the possibility of executing independent instructions is dependent on the availability of computing resources.[1] Furthermore, a criterion must be defined to assign instructions to *processors*. Depending on the model of parallel computation (at least among those we will discuss in this book) instructions are assigned according to one of the two following alternatives:

1. Each graph node (i.e. each instruction) is assigned to a different processor. In this case, arcs entering (leaving) a node can be viewed as the connections along which the processor receives (communicates) data.
2. Different paths (not necessarily independent ones) are assigned to different processors. In this case, instructions on the same path are executed by the same processor and constitute a sequential computation. Communication between processors takes place at those nodes where precedence relations are defined among nodes of different paths (see Figure 3.4).

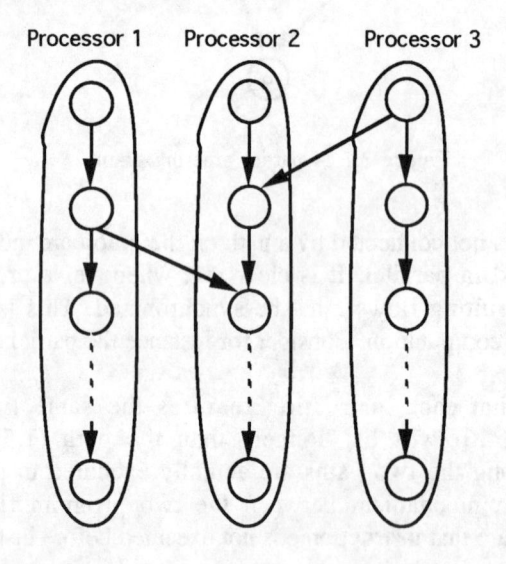

Figure 3.4 Assigning instructions to processors.

The graph shown in Figure 3.2 has been constructed based on simple observations on the nature of the computations generated by the program calculating the product of matrices. One can naturally ask at this point if there is a systematic way to obtain, starting from a given program, a partial ordering graph of the instructions, and, also, how to determine a graph that maximizes the potential parallelism of a sequential computation. Note, however, that it is not always possible to obtain efficient parallel algorithms

---

[1] In the following we will use the term processor with the (informal) meaning of a set of computational resources able to execute one or more instructions of a given program.

from efficient sequential algorithms (see Chapter 4). The problem of parallelizing sequential algorithms is an important one when studying compilers for parallel or vector machines.

At the beginning of this section we remarked that most programming languages correspond to an imperative view of computations. A conventional computer instruction, however, can be viewed both as a command to the processor, and as a function. The parameters of this function are memory locations (or the contents of these locations), and its results are also memory locations (or the values stored in these locations). The fact that an instruction of type I_j uses the values of k_j memory locations, and modifies the contents of h_j memory locations, can be written as:

$$I_j : M^{k_j} \to M^{h_j}$$

where M is the set of memory locations. In the RAM model (where memory locations are represented by registers) one has:

load: $M \to M$ (with direct addressing)

and:

add: $M^2 \to M$

In view of this, the intuitive notion stating that "independent instructions can be executed in parallel" can be formalized as follows. Let $D(I) \subseteq M$ denote the set of memory locations the values of which are used by some instruction I. Let $R(I) \subseteq M$ denote the set of locations the contents of which are modified by I. If I is of type I_j the number of elements in the sets $D(I)$ and $R(I)$ is k_j and h_j respectively. One has, for instance, $D(\text{load } r_1) = \{r_1\}$ and $R(\text{load } r_1) = \{A\}$. It can be shown that in order for two operations I and J to be independent it is sufficient that the following relations (called Bernstein conditions) be satisfied:

$R(I) \cap R(J) = \varnothing$
$D(I) \cap R(J) = \varnothing$
$R(I) \cap D(J) = \varnothing$

When the Bernstein conditions are not satisfied, there exists a logic dependence between I and J, and a sequencing must be performed in order to have a deterministic computation. At this stage, a way to determine the partial ordering graph that maximizes the potential parallelism of a sequential computation can be given. The Bernstein conditions are applied to each pair of instructions, and a precedence relation is introduced, i.e. an arc on the graph, if and only if at least one of them is not satisfied. As an example, one can verify that the graph in Figure 3.2 describes a maximally parallelized computation starting from the program in Figure 3.1.[2]

Since it describes possible parallel computations, a partial ordering graph is already a very abstract computational model. As noted before, each node could represent a processor, and each arc could represent a data communication along some physical connection. In this case, the number of nodes gives a measure of the hardware resource required by the computation, while the length of the longest path measures "parallel time".

We describe in the following a model which has an obvious correspondence to the study, conducted starting from the Bernstein conditions, of the potential parallelism of algorithms. This model is less precise, in a *formal* sense, than the models we will discuss in the remainder of this chapter. It is however well suited to the study of algorithm performance. For this reason we will refer to it as the *algorithmic model*. The model is described by the following rules:

1. at any instant, any number of processors can be used;
2. every processor can execute any operation (arithmetical or logical) in a unit of time;
3. data access is at no cost;
4. communication among processors is at no cost.

Much literature on parallel computation, especially in the 70s, referred to this model. The crucial resources needed by the model are the *number of parallel steps* (number of time units) and the *number of processors* used, as a function of problem size.

Example 3.3
One can verify that rules 1-4 above defining the algorithmic model are implied in the following "algorithm" to calculate the product of two matrices and to analyze the corresponding cost.

   Input: Two matrices, $A = (a_{ij})$ and $B = (b_{ij})$ of order n.

   Output: A matrix $C = (c_{ij})$ of order n such that $C = AB$.

   Steps:
   1. Compute in parallel the n^3 products $t_{ikj} = a_{ik} * b_{kj}$.
   2. Compute in parallel the n^2 sums
      $c_{ij} = t_{i\,1j} + t_{i\,2j} + ... + t_{inj}$, $i, j = 1, ..., n$,

---

[2]Note however that, using a different program, it is possible to obtain a computation characterized by a parallel time proportional to log n rather than to n (see Section 3.5.1).

using the scheme in Figure 3.5 (where $n = 8$ and $t_k = t_{ikj}$).

Cost: Step 1 requires one time unit and n^3 processors.
Step 2 requires $\lceil \log n \rceil$ steps and $n^2 \lfloor n/2 \rfloor$ processors.
In total $\lceil \log n \rceil + 1$ steps and n^3 processors are required.

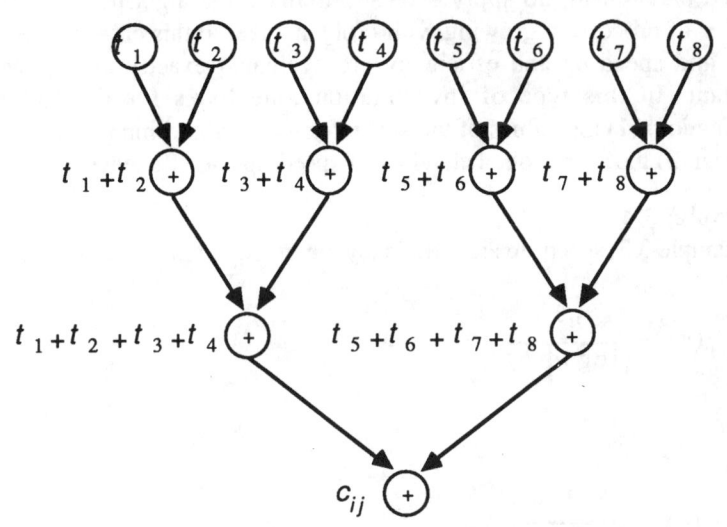

Figure 3.5 Computing the sum of eight numbers.

When the number p of processors is a constant (*limited parallelism*), algorithm performance is usually analyzed by means of two other measures: *speed-up* and *efficiency*. The speed-up $S_p(n)$ is defined as the ratio between the time cost of the best available sequential algorithm and the time cost of the parallel algorithm to be analyzed, when the latter has at most p available processors.

$$S_p(n) = \frac{T_1(n)}{T_p(n)}$$

where n is the problem size. It is easy to convince oneself that the best condition, difficult to achieve in practice, is $S_p(n) = p$. Ideally, we would like a problem that can be solved in a sequential time $T(n)$ to be solved in a parallel time $T(n)/p$, where p is the number of available processors. Generally speaking, a condition of this type cannot be achieved: in a parallel computation some time must unavoidably be used to coordinate processor activity (see Chapter 5). Moreover, parallelism could be limited by the nature of the problem itself.

The efficiency $E_p(n)$ is the ratio between speed-up and number of processors.

$$E_p(n) = \frac{S_p(n)}{p}$$

Obviously, $E_p(n) \leq 1$ and the best efficiency is when the equal sign applies.

The concepts of speed-up and efficiency, introduced while discussing limited parallelism, do apply even when analyzing algorithms that use a number of processors growing with problem size. In this case, one sets $p = p(n)$, and speed-up and efficiency are evaluated exactly as if p were a constant. In this type of investigation, one looks for the following information: (1) the values of the size n for which algorithm performance is best and (2) the asymptotic behaviour of speed-up and efficiency.

Example 3.4
In Example 3.3, speed-up and efficiency are given by:

$$S_{n^3}(n) = \frac{n^3}{\lceil \log n \rceil + 1}$$

and:

$$E_{n^3}(n) = \frac{1}{\lceil \log n \rceil + 1}$$

Note that efficiency vanishes in the limit $n \to \infty$. This means that the greater n, the longer the time during which most processors are inactive.

The foregoing suggests two possible ways to study parallel algorithms. One is to think of the number p of available processing units as a constant, and to try to synthesize parallel algorithms with an execution time $T_p(n)$ maximizing *speed-up*. On the other hand, p could be thought of as a function of problem size, that is, $p = P(n)$, and one could try to determine algorithms requiring the least possible parallel time.

The limited parallelism approach is best suited to parallel machines with a low number of processors. The case with no limits on parallelism is more stimulating from a conceptual point of view, since one can study, apart from "physical" constraints, the real parallel computational complexity of problems.

3.3 Boolean circuits

Boolean circuits were already widely in use (possibly in a more primitive form) when studying the computational complexity of Boolean functions long before concrete studies on parallelism had begun. Even "ordinary" computers are however parallel machines when observed in detail. Each computer is, in fact, made of electronic circuits with millions of switching elements and transmission lines operating in parallel. Boolean circuits are a logic model of electronic circuits.

The model works on problems represented by sequences of bits. For this reason, Boolean circuits characterize themselves as "string processors", and from this comes their general applicability. All information entering and exiting a real computer is encoded in a sequence of binary digits. Even if bit sequences represent data with some sort of structure, the meaning of a sequence depends on *interpretation*, which is external to the machine (see Example 2.12). Since operations are executed at the bit level, when an algorithm is expressed as a Boolean circuit its cost is called *bit cost* (while its complexity is called *bit complexity*). This is, as we shall see, a measure of *parallel cost*.

In a sense, the above characteristics make the Boolean circuit model the basis for the study of the complexity of parallel computations. As in the sequential case, however, many problems (and algorithms) are studied on models with a high level of abstraction. Determining the complexity relations between abstract models and circuits is therefore very important. There are, on the other hand, important problems that can be studied directly using Boolean circuits. Note in particular the following instances:

- complexity of elementary functions (arithmetic operations for instance);
- complexity of combinatorial problems, or any problem defined on finite structures (which can be therefore represented within a machine as strings on the alphabet {0,1});
- language recognition.

3.3.1 Definitions

The somewhat complex definition of a Boolean circuit uses the graph notion discussed in Section 2.3.

Definition 3.5
A Boolean circuit α, with n inputs and m outputs, is a *direct, labelled and acyclic graph*, $\alpha = (V,E)$. The elements of V (nodes) are numbered from 1 to $|V|$, and subdivided into four disjoint sets: (1) *input* nodes, (2) *constant* nodes, (3) *operation* nodes and (4) *output* nodes.

- The n input nodes do not have arcs entering them. Each one is labelled by a different variable symbol. In the following we will assume that these nodes are numbered from 1 to n, and that the label of node i is X_i, $i = 1, ..., n$.

- The constant nodes do not have arcs entering them. Each one is labelled by *0* or *1*.[3] The set of constant nodes can be empty. These nodes, if present, will be numbered from $n + 1$ to $n + c$ ($c \geq 1$).
- The m output nodes do not have arcs exiting them, and have exactly one entering arc. Each one is labelled by a different variable symbol. We will assume that these nodes are numbered from $|V| - m + 1$ to $|V|$, and that the label of node i is Y_i, $i = |V| - m + 1, ..., |V|$.
- Operation nodes have both entering and exiting arcs. Each one is labelled by the symbol of a Boolean function with one or more arguments, the number of arguments being given by the number of arcs entering the node. We will assume in the following that operation nodes are numbered from $n + c + 1$ to $|V| - m$, and that the label on node i, with j_i entering arcs, is $f_{l(i)}$, where $f_{l(i)}: \{0,1\}^{j_i} \to \{0,1\}$, and $i = n + c + 1, ..., |V| - m$. The set $F = \{f_1, ..., f_r\}$ of all functions labelling operation nodes of α is called the *basis* of the circuit. As a particular case, the set of operation nodes can be empty.

The number of arcs entering (exiting) a given node is called *fan-in* (*fan-out*) of that node. The maximum fan-in, over all nodes, is the circuit fan-in, and is determined by the basis functions. We will only consider a basis, called *canonical basis*, consisting of the logical sum, product and complement operations, which we will denote by the symbols $\vee$, $\wedge$ and $\neg$ respectively. It is known that the canonical basis is a complete one, in the sense that any Boolean function can be expressed by means of the logical sum, product and complement operations only. The fan-in for circuits defined on the canonical basis is obviously 2.

Definition 3.6

Let $\alpha = (V,E)$ be a Boolean circuit with n inputs and m outputs. The *size* of α, denoted by $Z(\alpha)$, is the total number of nodes in the circuit, that is, $Z(\alpha) = |V|$. It is $Z(\alpha) \geq n + m$. The *depth* of α, denoted by $D(\alpha)$, is the maximum length among all circuit paths (i.e. among all paths connecting input nodes to output nodes).

As we shall see, size and depth are the most important measures of complexity for Boolean circuits.

Figure 3.6 shows a simple Boolean circuit with four inputs and one output. Its size and depth are, respectively, 15 and 6.

---

[3] We use italics to distinguish the use of the symbols 0 and 1 as labels or as numerical values.

Boolean circuits 71

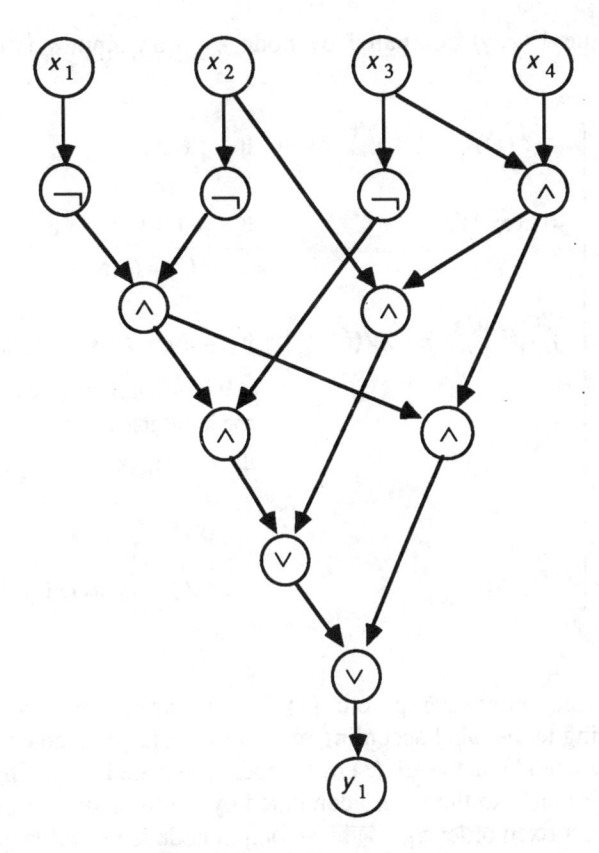

Figure 3.6 A Boolean circuit.

In the following we will assume that each input node is connected to an output node through a path. This is equivalent to stating that every input is used in determining the output values (in a sense that we will shortly illustrate). This *non-redundancy principle* bears consequences on the depth of a circuit. In fact, one can easily verify that a circuit α with n inputs and m outputs, with $n > m$, has a depth not less than $\lceil \log_t (n/m) \rceil + 1$, where t is the circuit fan-in.

The computation of a function can be associated with every circuit. To do this, we must introduce some preliminary notions. Let α be a Boolean circuit with n inputs and m outputs. Let Λ denote the labelling function of the nodes of α. It is then $\Lambda:\{1, ..., Z(\alpha)\} \to \{X_1, ..., X_n\} \cup \{Y_1, ..., Y\} \cup \{0,1\} \cup \{f_1, ..., f_r\}$. Define an *input* or a circuit α with n inputs, by assigning values of the set $\{0,1\}$ to the variables labelling the input nodes. Formally, an input γ is a function defined on $\{X_1, ..., X_n\}$ with values in $\{0,1\}$, $\gamma:\{X_1, ..., X_n\} \to \{0,1\}$. Assume further that labels *0* and *1* denote the values 0 and 1 respectively.

The value $V(k, \gamma)$ computed by node k on an input γ is defined as follows:

$$V(k, \gamma) = \begin{cases} \gamma(\Lambda(k)) & \text{if } 1 \leq k \leq n, \\ \\ 0 \text{ (resp. 1)} & \text{if } n+1 \leq k \leq n+c \\ & \text{and } \Lambda(k) = 0 \text{ (resp. } 1), \\ \\ f_{l(k)}(V(i_{j_1},\gamma), ...,V(i_{j_k},\gamma)) & \text{if } n+c+1 \leq k \leq Z(\alpha) - m, \quad (1) \\ & f_{l(k)} \text{ is the label on node } k, \\ & \text{and in the graph there are} \\ & \text{the arcs } (i_{j_1},k), ..., (i_{j_k},k), \\ \\ V(j, \gamma) & \text{if } k > Z(\alpha) - m \text{ there is the arc } (j, k). \end{cases}$$

In words, the value computed: (1) by an input node is the value corresponding to the label according to the input γ, (2) by a constant node is the value denoted by the label, (3) by an operation node is found by applying the (function) label to the values computed by the immediate predecessors, according to a fixed ordering, (4) by an output node is the value given by its only immediate predecessor.

Definition 3.7
The value computed by a circuit α with m outputs, on an input γ, is the sequence $Y(\alpha,\gamma) \in \{0,1\}^m$ computed in an ordered way by the output nodes $Z(\alpha) - m + 1, ..., Z(\alpha)$, that is, $Y(\alpha,\gamma) = V(Z(\alpha) - m + 1, \gamma) ...V(Z(\alpha), \gamma)$.

Definition 3.8
A circuit α with n inputs and m outputs computes the function $f:\{0,1\}^n \to \{0,1\}^m$ if, for any input γ, it is $Y(\alpha,\gamma) = f(X_1, ..., X_n)$.

Note that a circuit computes one and only one function (that is, it has a deterministic "behaviour").

Example 3.9 (One-bit adder)
Figure 3.7 shows a circuit computing the function $f(X_1, X_2, X_3) = <Y_1, Y_2>$, where $Y_1 = f_1(X_1, X_2, X_3) = (X_1 + X_2 + X_3) \text{ div } 2$ and $Y_2 = f_2(X_1, X_2, X_3) = (X_1 + X_2 + X_3) \text{ mod } 2$. The circuit consists of a subcircuit, shown in Figure 3.8, computing the exclusive OR of two bits. To make sketches more readable, we will use the following graphical conventions:

- edges will substitute arrows, assuming that the orientation is always from top to bottom;
- only one edge exits each node, while possible contact points are used to branch the edge to reach successor nodes. It is important not to evaluate incorrectly node fan-out. For instance, the fan-out of all input nodes in Figure 3.7 is two, and not one.

According to the discussion in Section 2.3, a Boolean circuit can be in general viewed as a problem solver. In this case, both instances and solutions are given by binary sequences. Note that, given its fixed configuration, a circuit can solve only instances with some given size. If size n instances of a problem R (denoted as usual by R_n) are described by $g(n)$ bits, and if solutions to instances of R_n are described by $h(n)$ bits, a Boolean circuit α solving R_n must have $g(n)$ inputs and $h(n)$ outputs.

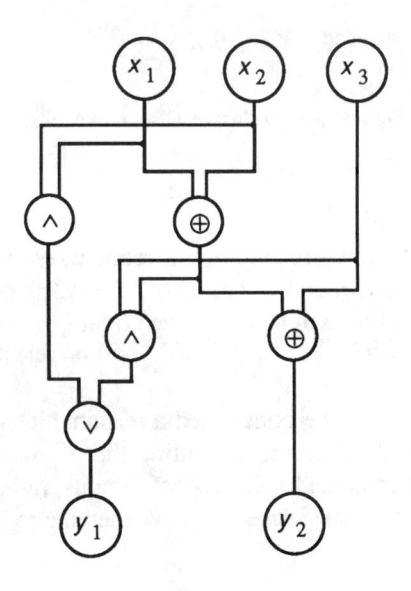

Figure 3.7 One-bit adder.

Definition 3.10
Let $R_n \subseteq \{0,1\}^{g(n)} \times \{0,1\}^{h(n)}$. A circuit α with $g(n)$ inputs and $h(n)$ outputs solves the problem defined by R_n if it computes the function $f_n: \{0,1\}^{g(n)} \to \{0,1\}^{h(n)}$, and if, whenever X encodes an instance of R_n and X has (at least) one solution, one has $(X, f_n(X)) \in R_n$.

Circuits with a single output can be used for language recognition on the alphabet $\Sigma = \{0,1\}$. More precisely, if A is a language on Σ (that is, $A \subseteq \Sigma^*$), let A^n denote the language $A \cap \Sigma^n$, consisting of the strings of A of length exactly n. A circuit α with n inputs, $x_1, ..., x_n$, and a single output, y, will therefore be said to recognize the language A^n, if $y = 1$, if and only if $x_1, ..., x_n \in A^n$.

(a) Circuit computing $Y = X_1 \oplus X_2$

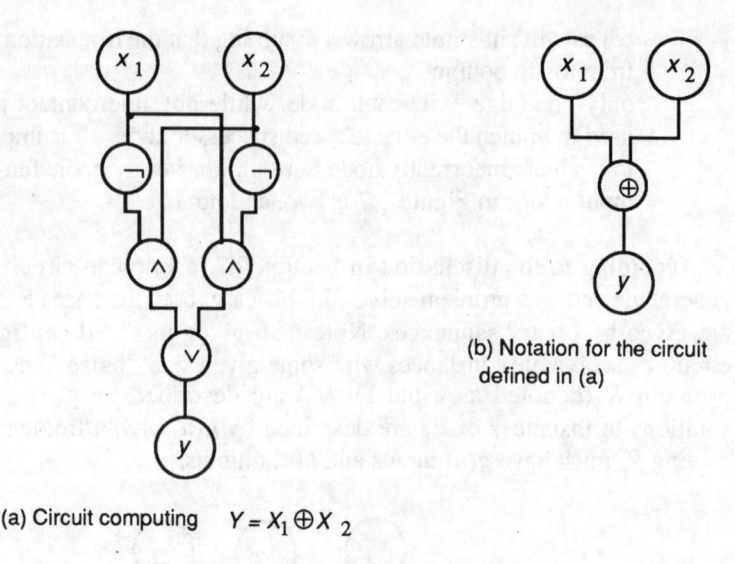

(b) Notation for the circuit defined in (a)

Figure 3.8 Exclusive OR of two bits.

Example 3.11
The Boolean circuit shown in Figure 3.6 recognizes strings $w \in \{0,1\}^4 \cap 0^*1^*$, that is, strings of length 4 where all 0s, if any, precede 1s, if any. For instance, the string "0011" will be recognized (that is, the circuit will give the output value $y_1 = 1$), while the string "0101" will be rejected (that is, $y_1 = 0$).

A single circuit cannot be considered a reasonable computation model. The reason is its fixed structure, preventing the solution of problems with any size. We must then consider *families* of circuits, rather than single ones, one for every possible instance size. We have then the following definition.

Definition 3.12
A family of circuits $\{\alpha_n\}_{n \in N}$ solves a problem R if and only if, for any $n \in N$, α_n solves R_n.

The concepts of size and depth can be immediately extended to families of circuits. The size (depth) of a family $\{\alpha_n\}_{n \in N}$ is a function $Z_\alpha : N \to N$ ($D_\alpha : N \to N$) such that α_n has size $Z_\alpha(n)$ (depth $D_\alpha(n)$). Whenever there is no possible ambiguity we will simply write $Z(n)$ and $D(n)$, omitting the circuit specification.

Example 3.13 (*n*-bit adder)
The sum of two *n*-bit integer numbers $X = x_{n-1}, ..., x_0$ and $Y = y_{n-1}, ..., y_0$ can be computed by the circuit shown in Figure 3.9, where each block labelled OBFA (standing for One-Bit Full Adder) is the "operation part" (that is, the operation nodes only) of a one-bit adder. Note that, in this case, input nodes are labelled by x_i and y_i, while the output ones are labelled by s_i, $i = 0, ..., n - 1$, and c. The arc entering s_i, $i = 0, ..., n - 1$, comes from the second exclusive OR of the corresponding OBFA (in other words, it is the input to the node labelled Y_2 in Figure 3.7), while the rightmost arc entering each OBFA module, except the first, comes from the OR node of the preceding OBFA (that is, the input to the node labelled Y_1 in Figure 3.7). Since *n* is arbitrary, the circuit shown in Figure 3.9 represents a family with size and depth $16n + 2$ and $2n + 5$ respectively.

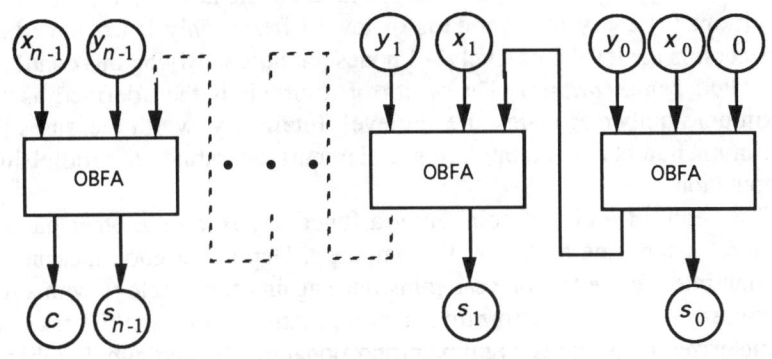

Figure 3.9 *n*- bit adder.

We will end this section with some remarks on the "intuitive" meaning of size and depth as measures of complexity for circuits. More precise ideas will be discussed in the following section.

Definition 3.8 (and, as a consequence, all definitions that make use of 3.8) gives a "meaning" to a Boolean circuit according to a functional view of semantics (see Section 2.2.1). From an operational point of view, a Boolean circuit computing a function *f* describes, in a formal way and on a more abstract[4] level, the computations done by an algorithm that computes *f* on a real computer. In any real machine computation it is possible, at least in principle, to identify the active computational elements, and their order of activation. With this "information" it is possible to construct a graph describing the computation in the following way: (1) each computational element is represented by a node; (2) if, while computing, a computational element *x* gives an output datum to an element *y*, the arc (*x*, *y*) is added to the graph.

---

[4]In a real computer, the operations we associate to the nodes of a Boolean circuit are, although simple, implemented with even more elementary functions.

A Boolean circuit is basically an abstraction of the computation graph constructed as discussed above. From this point of view, the operation nodes are computational elements, while arcs are connections for data transfer. Input and output nodes describe data transfer from and to the outside world respectively.

According to this operational view, the size of a circuit appears to be an appropriate estimate of the hardware cost of its computations. The size of a Boolean circuit is in fact strictly related to the number of computational elements in the corresponding graph. On the other hand, one could object that size is in general a pessimistic cost estimate. This is due to the fact that Boolean circuits are acyclic, making it impossible to use the same node more than once. An alternative hardware measure is circuit *width*. To define this resource, assume that all nodes in the circuit can be subdivided into levels in the following way: (1) input nodes and constant nodes are level 0 by definition; (2) every other node is of level i if and only if its immediate predecessors are level 0 or level $i-1$ nodes (circuits satisfying this condition are called *synchronous*). The width of a circuit is then defined as the maximum number of nodes at each level. Intuitively, width measures the maximum number of computational elements operating in parallel in a computation.

The depth of a circuit computing a function f is, on the other hand, a measure of the time sufficient to compute f. Depth is a good measure of parallel time. To see this one can think once again of the analogy with a real computer. A circuit performing a computation will, in this case, be characterized by a processing time proportional to the maximum number of elements crossed by a signal (since the operating time of a computational element is different from zero[5]), that is, to the depth of the circuit itself. Note now that time is really "parallel time", since the activity of nodes on paths, linking inputs to output, takes place in parallel.

3.3.2 Uniform circuits and complexity

Starting from the complexity measures we have identified, let us define a few important classes of problems in parallel computation theory.

Definition 3.14
Let $S:N \to N$. DEPTH($S(n)$) is the class of problems which can be solved by families of circuits of depth $O(S(n))$:

$$\text{DEPTH}(S(n)) = \{R \mid \exists \{\alpha_n\}_{n \in N}, \alpha_n \text{ solves } R_n \text{ and } D_\alpha(n) = O(S(n))\}$$

---

[5] We assume a null delay in data transmission along the connections (arcs) of the circuit (see Chapter 6).

Definition 3.15
Let $T:N \to N$. SIZE($T(n)$) is the class of problems which can be solved by families of circuits of size $O(T(n))$:

SIZE($T(n)$)={$R \mid \exists \{\alpha_n\}_{n \in N}$, α_n solves R_n, and $Z_\alpha(n) = O(T(n))$}

The problem of adding two n-bit integer numbers is simultaneously in both classes DEPTH(n) and SIZE(n), as shown by the family in Example 3.13.

Definitions 3.14 and 3.15 pose some difficulties, since, as will be discussed in the following example, they contain problems without algorithmic solution.

Example 3.16
Consider an arbitrary, non-recursive, subset W of the natural numbers. Since W is non-recursive, there is no algorithm such that, for any $x \in N$, it can decide whether $x \in N$ or $x \notin N$ in a finite time. Define then a function $w_n:\{0,1\}^n \to \{0,1\}$ (where $g(n) = n$ and $h(n) = 1$):

$$w_n(X) = \begin{cases} 1 & \text{If the number of bits of the binary representation of } n \text{ is an element of W;} \\ 0 & \text{Otherwise.} \end{cases}$$

There obviously exists a family which computes the function $w = \langle w_n \rangle_{n \in N}$. The nth circuit of the family must in fact always give either 0 or 1 for any $X \in \{0,1\}^n$, and circuits of this kind are easy to construct. We can even give size and depth of the family[6] (see Figure 3.10, where the label * denotes any operation with two Boolean arguments): $Z(n) = O(n)$ and $D(n) = O(\log n)$. In spite of this, since W is non-recursive, we cannot tell, for any $n \in N$, which one is the nth circuit of the family.

The above difficulties arise due to the fact that, by Definitions 3.14 and 3.15, the only requirement a problem R must satisfy in order to belong to the class DEPTH($S(n)$) (SIZE($T(n)$)) is the existence of a family of depth $O(S(n))$ (size $O(T(n))$) solving R. Note that, still by Definitions 3.14 and 3.15, no relation between circuits belonging to the same family is required. Apparently, though, it is not permissible that the structure of the nth circuit of the family be independent of the structure of the $(n + 1)$th circuit. If this were

---

[6]The value $O(\log n)$ of the depth depends on the input non-redundancy principle (see Section 3.3.1). If the principle did not apply, the computation of w could even be done by constant depth families.

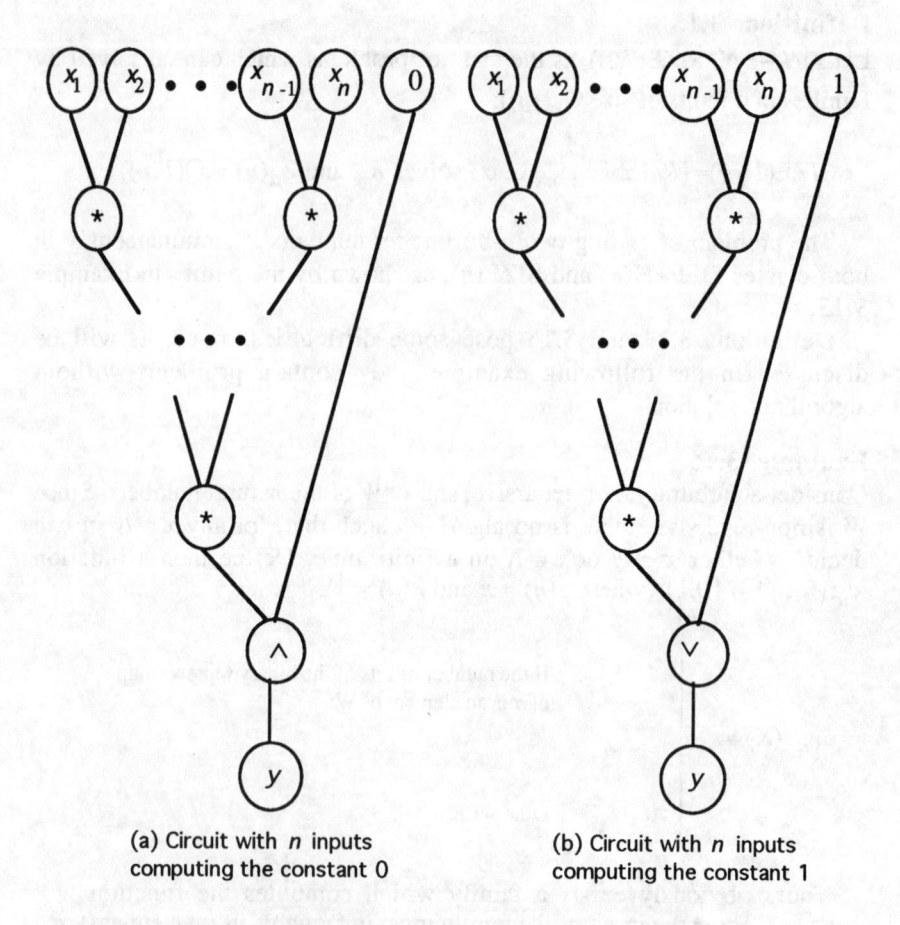

(a) Circuit with n inputs computing the constant 0

(b) Circuit with n inputs computing the constant 1

Figure 3.10 Circuits with depth $O(\log n)$ computing constants.

the case, the family could not be described using a finite amount of information.

To "restrict" classes in such a way that they contain only problems amenable to an *algorithmic* solution (which can therefore be described by a finite amount of information), we must impose a *uniformity* condition on the circuits in a family.

Definition 3.17
A family of circuits $\{\alpha_n\}_{n \in N}$ is *uniform* if and only if there is an algorithm which, for a given n, generates a description of α_n.

The result given by the algorithm will be a "linear" description of the circuit. A possible linear description is simply a listing of the nodes, where for each node its number, label and the number of immediate predecessor nodes are specified. Since we limit ourselves to circuits with fan-in 2 (as a consequence of adopting the canonical basis), each node k in the listing can be represented by the quadruplet (k, t, k_1, k_2), where t is the label and (k_1, k) and (k_2, k) are arcs in the circuit (if the number of arcs entering k is less than 2 the value of k_1 and/or k_2 is a special symbol, for instance).

Various ways to specialize the notion of uniformity have been proposed in the literature (see the paper by Ruzzo [Ru81]). We will briefly discuss only two of these. The fundamental requirement in both cases is that the complexity of the generating circuit be not too high compared to that of the generated circuit. Often, deterministic Turing machines are chosen as generators. In this way, the cost of producing a circuit description can be accurately given in terms of the quantity of tape used and/or the number of steps in the computations.

Definition 3.18
A family of circuits $\{\alpha_n\}_{n \in N}$ of size $T(n)$ is *log space-uniform* if and only if there exists a deterministic Turing machine (TM) of space complexity $O(\log T(n))$ which, on input 1^n, generates a description of α_n.

Note that the input to the TM is presented in unary notation. In this way n is the length of the input, rather than its value (in which case the length would obviously be $\lceil \log n \rceil$).

Definition 3.19
A family of circuits $\{\alpha_n\}_{n \in N}$ of size $T(n)$ is *P-uniform* if and only if there exists a deterministic TM of time complexity $T(n)^{O(1)}$ which, on input 1^n, generates a description of α_n.

It is easy to realize that every log space-uniform family is also P-uniform. In fact, a TM using $O(\log T(n))$ cells on its working tapes cannot perform more than $T(n)^{O(1)}$ steps, unless it enters an infinite loop (see Proposition 2.60).

Example 3.20 (uniformity of the family in Example 3.13)
The uniformity of the family which solves the problem of adding integer numbers can be proved in the following way. Assume the generation of the generic nth circuit is from right to left. Note first of all that the number of OBFA modules to be generated is n, and therefore $O(\log n)$ bits are sufficient to keep track of the generation process. Furthermore, since the size is $Z(n) = 16n + 2$, $O(\log n)$ bits are sufficient to record the number of a node. The number of nodes within a single OBFA is a constant. The relative quadruplets can then be generated in space proportional to $\log n$. Finally, when going from generating a module to generating the next one, say from the jth to the $(j + 1)$th, the only information to be stored is the number of the node calculating the jth carryover. This uses again space $O(\log n)$. A

working space greater than $O(\log n)$ will never be used. Since $Z(n) = 16n + 2$, the family being discussed is log space-uniform.

Once the notion of uniformity has been introduced, Definitions 3.14 and 3.15 can be modified by adding new complexity classes.

Definition 3.21
Let $S:N \to N$. $\text{DEPTH}_L(S(n))$ ($\text{DEPTH}_P(S(n))$) is the class of problems which can be solved by log space-uniform (P-uniform) circuit families of depth $O(S(n))$.

Definition 3.22
Let $T:N \to N$. $\text{SIZE}_L(T(n))$ ($\text{SIZE}_P(T(n))$) is the class of problems which can be solved by log space-uniform (P-uniform) circuit families of size $O(T(n))$.

It is interesting to note that when studying lower complexity bounds it is not very important to require the uniformity condition on circuit families. On the contrary, according to Definitions 3.13, 3.14, 3.21 and 3.22, one has $\text{DEPTH}_*(S(n)) \subseteq \text{DEPTH}(S(n))$ and $\text{SIZE}_*(T(n)) \subseteq \text{SIZE}(T(n))$ (where * denotes any uniformity criterion). If, then, there is a lower bound on a problem with respect to the non-uniform model, the same bound will even more strongly apply to a model with some uniformity criterion. It follows then that a lower bound for a non-uniform model is a very strong condition.

Lower bounds on the computational complexity of some classes of functions can be obtained by means of simple arguments based on circuit fan-in. Consider a function $f(x_1, x_2, ..., x_n) = x_1 @ x_2 @ ... @ x_n$, where @ is any Boolean associative operator. Since the value of $f(x_1, x_2, ..., x_n)$ depends on all arguments, the minimum depth needed to "combine" the n inputs into a single output value is $\lceil \log_t n \rceil$ where t is the circuit fan-in. For instance, computing the logic sum, or the logic product, of n variables requires, for circuits defined on the canonical basis, a depth $\Omega(\log n)$. This value can be easily obtained from the circuit layout shown in Figure 3.5.

As suggested in Example 3.16, only constructive results can be accepted from the point of view of upper complexity bounds. For instance, to establish that a problem can be solved by a family of circuits of depth $S(n)$, it must be actually possible to construct the circuits. It is not sufficient to prove their existence.

In the case of uniform circuit families there are important results linking the size and depth resources to the sequential time and space resources, respectively.

Proposition 3.23 [Bo77]
If $S(n) \geq \log n$ is a space-constructible function (see Definition 2.56), then:

$$\text{DEPTH}_L(S(n)) \subseteq \text{SPACE}_{TM}(S(n))$$

and:

$$\text{NSPACE}_{TM}(S(n)) \subseteq \text{DEPTH}_L(S^2(n))$$

Proof. We will limit ourselves to proving the second relation only. This relation allows one to transfer to parallel computations positive results obtained for sequential complexity. The proof in [Bo77] concerns decision problems. Consider then a non-deterministic TM M of space complexity $S(n) \geq \log n$ solving a decision problem R. We must prove that the computation performed by M on inputs of length n can be simulated by a circuit of depth $O(S(n)^2)$. With the input $w = w_1 w_2 \ldots w_n$, $w_k \in \{0,1\}^n$, $k = 1, \ldots, n$, kept fixed, the number of possible configurations of M is $N = qnc^{S(n)}S(n)$, where q is the number of states of the TM control and c is the number of symbols in the alphabet of the working tape (see Proposition 2.60). Consider the graph G_M of the computation of M. Let 1 be the node corresponding to the initial configuration, and let $f_1, \ldots, f_r$, $r \leq N$, the nodes corresponding to the final configurations. The problem of deciding whether the instance R encoded by the input w has a YES solution (i.e., if M accepts w) is then coincident with the problem of establishing whether at least one of the nodes $f_1, \ldots, f_r$ can be reached (is accessible) from node 1. The circuit corresponding to M solves the latter problem by computing the transitive closure of the graph of the computation of M. If $X = (x_{ij})$ is the adjacency matrix of G_M, let $X^* = (x_{ij}^*)$ denote the adjacency matrix of the transitive closure of G_M. The circuit can be constructed in the following way. Consider two configurations i and j, and assume that in configuration i the head is on cell k of the input tape. There are four cases.

1. A transition from i to j is possible only if $w_k = 1$. In this case, bit w_k must be connected to the input x_{ij} of the (sub)circuit computing the transitive closure of X (see Figure 3.11(a)).
2. A transition from i to j is possible only if $w_k = 0$. In this case the value of w_k complemented is connected to x_{ij}.
3. A transition from i to j is always possible. In this case the value connected to x_{ij} is the constant 1.
4. A transition from i to j is never possible. In this case the value connected to x_{ij} is the constant 0.

Since the transitive closure of a graph with N nodes can be determined by computing the $(N-1)$th power of the relative adjacency matrix (see Section 2.5), the circuit computing the transitive closure of G_M can be obtained by assembling $\log(N-1)$ levels of circuits computing the product of matrices of type X^h, according to the scheme shown in Figure 3.11(b). Furthermore, the product of matrices of order N can be implemented by a circuit of depth $O(\log N)$ (see Section 3.5). It follows that the computation of $X^{N-1} = X^*$ can be implemented by a circuit of depth $O(\log^2 N)$. Finally, the output of the whole circuit is the logic sum of the elements of the matrix X^* describing paths between the initial configuration and the r final configurations. As noted before, this sum can be implemented by a circuit of depth $\log N$ (see also Section 3.5). The depth of the resulting circuit is therefore $O(\log^2 N) = O(S^2(n))$.

Concerning the uniformity of the circuit shown in Figure 3.11(a), note that circuits for matrix multiplication and for logic sum are uniform. The "suspicious" part is the determination of the input and output circuit connections for Boolean closure. It is however possible, with a careful enumeration of the configurations of M, to give a uniform description of the whole circuit. Note finally that the constraint that S must be space constructible is linked to the uniformity problem.

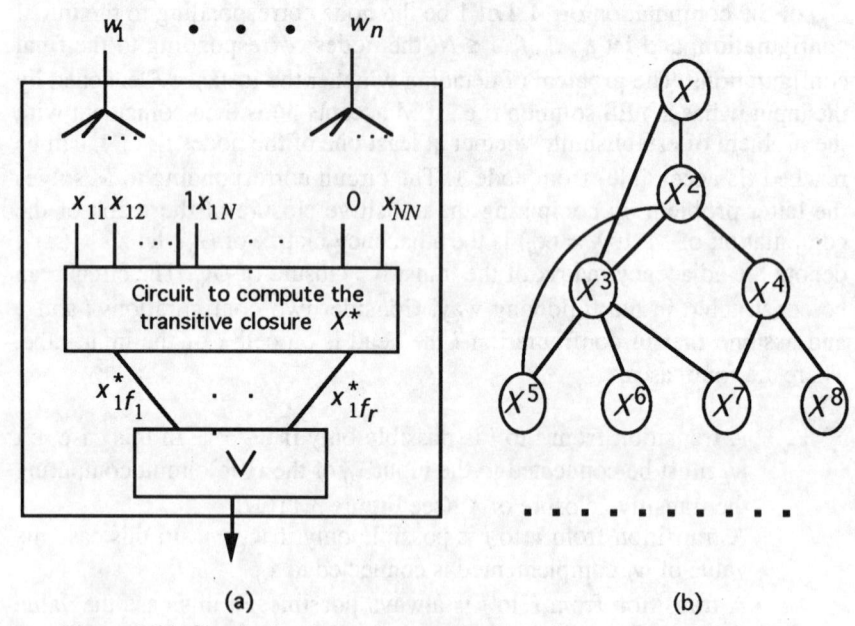

Figure 3.11 Circuit simulating the computations of a non-deterministic TM.

Proposition 3.23 can be extended to the general case of search problems (i.e. problems constructing a solution). Let $S(n) \geq \log n$ be a space constructible function, and let M be a TM of space complexity $S(n)$ computing a function $f = \langle f \rangle_{n \in N}$ implementing a problem R. According to the conventions it is $f_n:\{0,1\}^{g(n)} \to \{0,1\}^{h(n)}$. Assume further that $g(n) \geq n$ and $h(n) = n^{O(1)}$. Define the language A_f in the following way:

$A_f = \{(x, i) \mid \text{the } i\text{-th bit of } f(x) \text{ is } 1\}$

The pair (x, i) is presented to the machine as a sequence $\{0,1\}^{g(n)}\#1^i$. Starting from M it is possible to obtain a TM, say M', which accepts A_f using $O(S(n))$ cells of tape. Simply put, M' uses M to compute the bits of $f(x)$, without writing them to the output tape. When it arrives at the i th bit, if this is 1 it accepts the input, otherwise it rejects it. The amount of tape used by M' is that used by M plus the space needed to count the bits of the result, these being at most $h(n)$. According to the hypothesis, $h(n) = n^{O(1)}$, the space to count bits is $O(\log n)$, and since $S(n) \geq \log n$ the space request is not altered (in order of magnitude). Conversely, using a TM M recognizing A_f it is possible, with $h(n)$ calls of M, to compute all bits of $f(x)$. A circuit to compute f_n consists then of $h(n)$ copies, working in parallel, of the circuit corresponding to the machine M' according to Proposition 3.23.

The result expressed by Proposition 3.23 is one of the many manifestations of the thesis of parallel computation (see Section 3.1). This thesis characterizes problems, the solution of which can be considerably accelerated in parallel, as those, and only those, problems amenable to sequential solutions with a high space efficiency.

Proposition 3.24 [PF79]
If $\log T$ (and therefore T) is a space-constructible function, then

$$\text{SIZE}_L(T(n)) \subseteq \text{TIME}_{TM}(T(n)) \subseteq \text{SIZE}_L(T(n)\log T(n))$$

Proof. See [PF79].

Note that, generally speaking, Propositions 3.23 and 3.24 are not simultaneously valid. In other words, from the fact that a problem R is in $\text{SPACE}_{TM}(S(n))$ and in $\text{TIME}_{TM}(T(n))$, we cannot conclude that there exists a uniform family of circuits of depth $O(S^2(n))$ and size $O(T(n)\log T(n))$ solving R. We can only say that there is a family of depth $O(S^2(n))$ and a family (possibly different from the first one) of size $O(T(n) \log T(n))$ solving R. It is interesting that problems simultaneously solvable in time $T(n)$ and space $S(n)$ can be characterized, in parallel, in terms of the simultaneous resources size and width (rather than in terms of size and depth).

The equation:

$$\text{TIME-SPACE}_{TM}(n^{O(1)},(\log n)^{O(1)}) = \text{SIZE-WIDTH}_L(n^{O(1)},(\log n)^{O(1)})$$

has been in fact proved in [PF79], where $\text{SIZE-WIDTH}_L(T(n),S(n))$ denotes the class of problems which can be solved by log space-uniform circuits of size $T(n)$ and width $S(n)$.

The following two definitions introduce parallel complexity classes relative to simultaneous bounds on size and depth.

Definition 3.25
For any $k \geq 1$:

$$NC^k = \text{SIZE-DEPTH}_L(n^{O(1)}, \log^k n)$$

that is, NC^k is the class of problems which can be solved by families of log space-uniform circuits of depth $O(\log^k n)$ and size $n^{O(1)}$.

Definition 3.26

$$NC = \bigcup_{k \geq 1} NC^k$$

By substituting, in Definitions 3.25 and 3.26, the notion of log space-uniformity with the weaker notion of P-uniformity, one obtains the classes NC^k(P-uniform), containing problems solvable by P-uniform families of depth $O(\log^k n)$ and size $n^{O(1)}$, and NC(P-uniform) $= \bigcup_k NC^k$(P-uniform). It is not known whether, for any k, $NC^k = NC^k$(P-uniform). It has been proposed however that NC $\subset$ NC(P-uniform) (see [Co85]).

Many reasons make the class NC a particularly interesting one, both from a theoretical and a practical point of view. First of all, problems in NC possess very fast parallel algorithms. Secondly, it has been proven that the definition of NC is robust (that is, independent of the particular computation model). For instance, the class NC can also be characterized as the class of problems solvable by means of a parallel RAM (see Section 3.4) in polylogarithmic time and with a polynomial number of processors. Thirdly, it is thought that parallel machines with a polynomial limit to the number of computational elements (circuit nodes as well as simple processors) are in some way feasible, that is, they can be physically constructed. Lastly, by Proposition 3.23, problems in NC are also characterized by highly space-efficient sequential algorithms.

Most of the algorithms we will introduce in Chapter 4 belong to the class NC, or to the *arithmetic* NC class, which we will define in Section 3.5.1.[7] We anticipate below a brief list of very important computational problems. For each of them an NC algorithm has been found (for a more exhaustive list see [Co85]).

---

[7]Note that we indifferently speak of an NC problem and of an NC algorithm.

- Addition and multiplication of two n-bit integer numbers are in NC^1. It is clear that the circuit for addition is not the one shown in Figures 3.9 and 3.13, which has a depth linear in n.
- Iterated addition, that is, the sum of n n-bit numbers, is also in NC^1 [BCP83], while iterated multiplication is in NC^2. Note that this last result follows trivially from the result on the multiplication of two integers. Conversely, the result on iterated addition is not trivial.
- Division of two integers is in NC^1(P-uniform) [BCH84] and NC^2. It is not known whether it is also in NC^1. Reif [Re86] however proved that it is possible to construct uniform circuits for division with $O(\log n \, \log\log n)$ depth and polynomial hardware.
- Multiplication of two Boolean matrices is in NC^1, as one can immediately see by applying the results on the multiplication of two integers and the iterated addition. This result has been used in proving Proposition 3.23.
- The computation of the transitive closure of a graph and the accessibility problem in a graph are in NC^2 (see Proposition 3.23).
- Sorting n n-bit numbers is in NC^1.
- Recognizing context-free languages is in NC^2 [Ru81].

As in the sequential case, the concept of reduction plays an important part in the complexity of parallel computation. For the class NC, containing the fastest algorithms, the NC^1 *reducibility* is considered.

Definition 3.27
A problem R is NC^1 reducible to R', and one writes $R \leq_{NC} R'$, if and only if there exists a log space-uniform[8] family of circuits of depth $O(\log n)$ solving R and having an oracle for R'. An oracle for R' is a node with inputs $y_1, ..., y_r$ and outputs $z_1, ..., z_s$, the values of which satisfy the relation $((y_1, ..., y_r),(z_1, ..., z_s)) \in R'$. In the computation of the total depth of the circuit an oracle contributes a value $\lceil \log(r + s) \rceil$.

The relation $\leq_{NC}$ satisfies the following conditions (see Section 2.5 and [Co85]):

- if $R \leq_{NC} Q$ and $Q \leq_{NC} S$, then $R \leq_{NC} S$ (in other words, $\leq_{NC}$ is transitive);
- if $R \leq_{NC} Q$ and $Q \in NC^k$, then $R \in NC^k$.

---

[8]The definition of NC^1 reducibility in [Co85] refers to a uniformity condition stronger than log space uniformity (see also [Ru81]).

Example 3.28

The problem, which we will denote by DIV, of computing the first n bits of the quotient of two n-bit integer numbers is NC^1 reducible to the problem, which we will denote by REC, of computing the first n bits of the reciprocal of an n-bit number, that is, DIV $\leq_{NC}$ REC. Since multiplication of n-bit integer numbers is in NC^1, provided one has an oracle for REC, the circuit for DIV is simply the NC^1 multiplier, the inputs of which are the dividend and the reciprocal of the divisor.

3.4 Parallel Random Access Machine (PRAM)

A *parallel* RAM (PRAM in short) is an idealized multiprocessor model. The term stems from the fact that single processors are sequential RAMs (see Section 2.2.2) with access to a shared memory, used as a means of communicating. In Section 3.4.1 we will study a PRAM model closely related to the original model, introduced in [FW78]. Afterwards, in Section 3.4.2, we will discuss variations found in the literature. Finally, in Section 3.4.3, we will study relations among different PRAM models, and between PRAMs and uniform circuits.

3.4.1 Original model

A PRAM, as shown in Figure 3.12, is an infinite collection of processors, $P_0, P_1, ...$, each of which has access to a local memory. In Figure 3.12 the local memory of processor P_i is denoted by R_i. All processors can read from and write to a global memory. Local and global memories are made of an unlimited sequence of registers, each capable of containing an integer of arbitrary length. Global memory registers will be denoted by m_i, $i = 0, 1, ...$, while for local memory registers we will use the symbols r_i, $i = 0, 1, ...$.

Among the local registers of a single processor, three have a special meaning. We will use for these the special identifiers PC, A and F:

1. PC denotes the *program counter* register (used in an identical fashion as in RAMs).

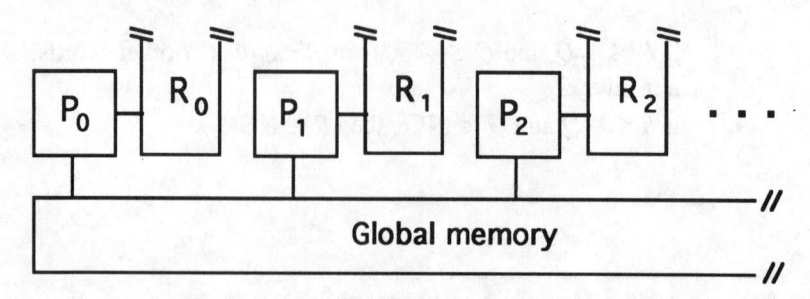

Figure 3.12 Scheme of a parallel RAM.

2. A indicates the *accumulator* register (also used in an identical fashion as in RAMs).
3. F denotes a *status indicator* register (flag). F contains 1 if the processor is active, 0 otherwise.

The PRAM model in [FW78] also contains read and write media, and, as a consequence, input/output instructions. According to more recent conventions, the model we present lacks these instructions. We assume instead that, at the start of each computation, the input is already present in global memory, and that when the execution is terminated results are read from specially designated global memory registers.

All processors execute only one program as in the RAM case. The program is an intrinsic part of the machine, and is not stored on a read/write medium. It consists of a sequence of instructions, possibly preceded by a label.

Table 3.1 shows the instruction set of our PRAM. Conventions needed to interpret the table are analogous to those in the RAM case. Keep however the following in mind:

1. A direct operand could also consist of an expression of the form mi meaning either (a) the register m$_i$ (if used in a STORE instruction) or (b) the contents of register m$_i$ (in all other cases).
2. The meaning of the indirect operand @ri, in an instruction I, is always dependent, as in the RAM case, on the contents of register r$_i$ of the processor executing I. Assuming that this register contains the number k, @ri could stand for either the register m$_k$ (if I is a store instruction) or the contents of m$_k$ (in all other cases).
3. The notation @mi is meaningless.

Note that points 2 and 3 represent limitations on the use of indirect addressing. Using a global register as a base, indirect addressing is not possible either to local or to global memory. Indirect addressing to global memory can only be done using a local register as a base. The choice is made on simplicity requirements (additional possibilities are usually considered in the literature).

We will describe the meaning of the instructions HALT and FORK. When a processor executes HALT the processor itself is halted (flag F takes the value 0, meaning inactive processor). The effect of the execution of a FORK *label* by a processor P$_i$ is the following:

- A currently inactive processor is found and activated. Let it be P$_j$.
- The content of the accumulator of P$_j$ is initialized with the content of the accumulator of P$_i$.

- The program counter of P_j is initialized with the value of *label* (the first instruction executed by P_j is then just the one preceded by *label*).

In the following, for definiteness, we will assume that the processor found and activated is the one with the lowest index (among the inactive ones).

A PRAM computation is defined as follows. The n input data, integer numbers of arbitrary length, are stored in the global memory registers m_0-m_{n-1}. Registers A, PC and F of processor P_0 are initialized with the values n, 1 and 1 respectively. All other registers, local and global, are zeroed: the only processor initially active is then P_0. The computation stops when any of the following conditions is satisfied:

1. Two or more processors attempt simultaneously to write to the same global memory location. In this case the result of the computation is undefined.
2. Processor P_0 executes HALT. In this case the result of the computation (i.e. the value computed by the machine) can be read in m successive global memory registers starting from a preset address.

Note that, even if the program is the same, at a given instant different processors can execute different instructions. We have in fact seen that each processor can be activated starting from any program instruction. Furthermore, one must consider the presence of conditional branching instructions, the result of which is in general different for each processor. This justifies the existence of a program counter register in each processor. In spite of this PRAMs are *synchronous*, and all processors initiate the execution of their respective instructions at the same instant.

Instruction	Meaning
LOAD operand	A ← R(operand)
STORE operand	L(operand) ← R(A)
ADD operand	A ← R(A)+R(operand)
SUB operand	A ← R(A)-R(operand)
JUMP label	PC ← R(label)
JZERO label	R(A)=0 ⇒ PC ← R(label)
JGTZ label	R(A)>0 ⇒ PC ← R(label)
FORK label	see text
HALT	F ← 0

Table 3.1 PRAM instruction set.

Definition 3.29
Let $g, h: N \to N \cup \{\perp\}$. A PRAM computes the function $f = \langle f_n \rangle_{n \in N}$, where $f_n: Z^{g(n)} \to Z^{h(n)} \cup \{\perp\}$, if, for any input $x \in Z^{g(n)}$:

- if $f_n(x)$ is defined, the computation stops when HALT is executed by P_0, and $h(n)$ successive global memory registers, starting from a preset address, contain in an ordered way the components of $f_n(x)$;
- if $f_n(x)$ is undefined, the computation diverges or it stops due to the attempt, by two or more processors, simultaneously to write to the same global memory register.

Definition 3.30
A PRAM solves the problem $R = \langle R_n \rangle_{n \in N}$, if and only if it computes a function $f = \langle f_n \rangle_{n \in N}$ implementing R.

The original PRAM model was introduced as a language recognition automaton, rather than as a problem solver. The conventions used about the beginning and the end of a computation were then slightly, but not substantially, different. In particular, since the result of a recognition is a binary datum (acceptance/refusal), this datum was read directly from the accumulator register.

Example 3.31 (program to compute the function $f(n) = \lceil \log n \rceil$)
The basic idea is as follows: an iterative scheme is used where, at each iteration, each active processor activates another one and assigns to it a progressive identification number. As soon as a processor receives the identification number n, the computation stops, and the number of completed iterations gives the desired value. In fact, since at each iteration the number of processors doubles itself, the identification number n will be assigned during the i th iteration, where i is the least integer such that $2^i \geq n$, and therefore $i = \lceil \log n \rceil$.

Figure 3.13 shows processor behaviour at different times. If the number k is present at the intersection of row i and column j, it means that processor P_i at the j th iteration activates processor P_k. If we let $k = p(i, j)$ we can convince ourselves that the function $p(i, j)$ is thus defined:

$$p(i, j) = i + 2^{j-1}, \quad j > \lceil \log i \rceil$$

The task of processor P_0 is to initialize the activation loop, and to compute, at every iteration, the value 2^{j-1} and to make it available to the other processors by means of a global memory register. Besides this, P_0 counts the number of iterations which, at the end of the computation, will be the result.

90   Parallel computation models

```
  j      1   2   3   4   5   6   . . .
i
  1      2   3   5   9  17  33   . . .
  2      -   4   6  10  18  34   . . .
  3      -   -   7  11  19  35   . . .
  4      -   -   8  12  20  36   . . .
  5      -   -   -  13  21  37   . . .
  6      -   -   -  14  22  38   . . .
  7      -   -   -  15  23  39   . . .
```

Figure 3.13 Activation scheme for Example 3.31.

The program, shown in Figure 3.14, is logically subdivided into two parts. The first one, shown in the leftmost column, is executed by processor P_0, the other one is executed by the remaining processors.

Each processor stores in the local register r_1 its own activation number. The processor given the number n signals the end condition by zeroing the global register m_1. Registers m_2 and m_3 contain, at the j th iteration, the values 2^{j-1} and j respectively. The result $\lceil \log n \rceil$ is then read from register m_3. By carefully analyzing the code in Figure 3.14 one finds that, when the computation stops (that is, when processor P_0 executes HALT), the program counters of all other processors contain the number of the instruction with the "sync" label. In our case, the actual instruction is unimportant, since the PRAM computation ends. The program however, besides computing the function $\lceil \log n \rceil$, can be seen as a utility subprogram to initialize $2^{\lceil \log n \rceil}$ processors to be used in a subsequent computation. It is therefore important to guarantee that the initialization phase is terminated in a well-defined state (see Example 3.33).

Example 3.32
With the unit cost criterion the computation of the function $f(n) = \lceil \log n \rceil$ can be executed by a PRAM in time $T(n) = O(\log n)$ using $P(n) = 2^{\lceil \log n \rceil}$ processors. Analyzing the program in Figure 3.14 it is clear that processor P_0 executes $\lceil \log n \rceil$ times the loop updating registers m_2 and m_3, checking the termination conditions. Since the number of instructions in the loop is a

```
1        load    #1              
2        fork    begin           
3        store   m1       16 begin  sub    m0
4        store   m2       17        jzero  end
5        nop              18        add    m0
6 loop0  load    m1       19        store  r1
7        jzero   end0     20 loop   load   m1
8        load    m3       21        jzero  sync
9        add     #1       22        load   r1
10       store   m3       23        add    m2
11       load    m2       24        fork   begin
12       add     m2       25        nop
13       store   m2       26        nop
14       jump    loop0    27        nop
15 end0  halt             28        jump   loop
                          29 end    store  m1
                          30        load   m0
                          31        store  r1
                          32        nop
                          33 sync   ...
```

Figure 3.14 PRAM code for Example 3.31.

constant, the program execution time is determined by the term $\log n$. The number of activated processors is $2^{\lceil \log n \rceil} + 1$.

Example 3.33 (computing the sum of n numbers)
The PRAM program shown in Figure 3.15 computes the sum of n numbers in time proportional to $\lceil \log n \rceil$. In writing the program we assumed the following:

1. The number $\lceil \log n \rceil$ is stored in register r_1 of processor P_0.
2. The $2^{\lceil \log n \rceil} + 1$ processors participating in the computation are initially active. The first instruction executed by P_0 is 1. The other processors start instead from instruction 15.
3. Processor P_i, $1 \leq i \leq 2^{\lceil \log n \rceil}$, has stored, in register r_1, a number j_i, between 1 and $2^{\lceil \log n \rceil}$, identifying it in a unique way.

The above hypotheses can be easily justified in view of Example 3.31. Generally speaking, regarding the time cost, one must consider an additional term $O(\log n)$, due to the initialization phase.

Figure 3.16 describes the contents of registers $m_1, ..., m_n$ during successive iterations. (In Figure 3.16 the special case $n = 8$ is considered. For the sake of simplicity, we assume that the numbers to be added are exactly 1, 2, ..., 8.) The computation stops after $\lceil \log n \rceil$ steps when P_0 executes HALT. Besides counting the number of iterations, at step k P_0

92   Parallel computation models

1		nop		15	load	r1
2		nop		16	sub	#1
3		nop		17	store	r0
4		load	#1	18	load	@r0
5		store	m0	19	store	@r1
6	loop0	load	r1	20	load	r1
7		jzero	end	21 loop	add	m0
8		sub	#1	22	store	r0
9		store	r1	23	load	@r0
10		load	m0	24	add	@r1
11		add	m0	25	store	@r1
12		store	m0	26	load	r1
13		jump	loop0	27	nop	
14	end	halt		28	jump	loop

Figure 3.15 PRAM program for Example 3.32.

computes the value $d = 2^k$, which is made available to the other processors by means of register m_0. The other processors start the computation by shifting the whole input a register forward, and thereby freeing register m_0. Therefore, at step k, processor P_i adds the value contained in m_{i+d}, $1 \le i \le 2^{\lceil \log n \rceil}$, to the contents of m_i. At the end of the computation the result can be read from register m_1.

The description of a PRAM as a parallel multiprocessor machine is basically a *model* description. From a formal point of view a PRAM can be defined as follows.

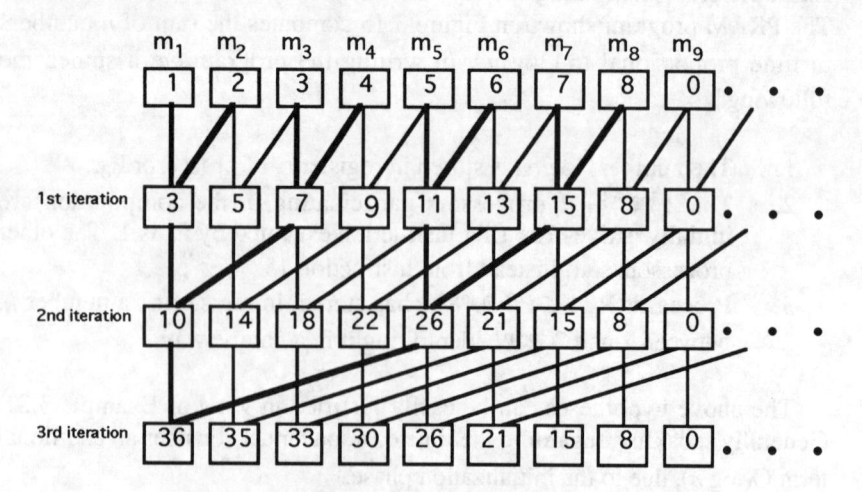

Figure 3.16 Parallel sum of eight numbers.

Definition 3.34
A PRAM is a pair (I, P), where I is a finite-length program, and $P:N \to N$ is a bound on the number of processors available in a computation (*processor bound*). Note that, if $P(n)$ can be described by a finite amount of information, the PRAM can then be described in a finite way.

A PRAM can be mapped into a family of Boolean circuits, rather than into a single circuit. This is due to the fact that there is no *a priori* constraint on the amount of computational resources which can be used by a PRAM. In other words, whereas a circuit has a fixed structure, a PRAM possesses a variable one. However, in the PRAM case we are also interested in a uniformity condition, which will obviously be relative to the program.

Definition 3.35
A PRAM is *uniform* if the finite-length program controlling the computation is the same for any possible input size.

The uniformity condition allows one to map a correspondence between PRAMs and uniform Boolean circuits. Due to the generality of the Boolean model, this correspondence gives one of the "strong" theoretical motivations to the introduction of the PRAM as a parallel computation model. The relations between uniform circuits and PRAMs will be briefly discussed at the end of Section 3.4.3.

Definition 3.36
$\text{TIME}_{\text{PRAM}}(T(n))$ is the class of problems solvable by a uniform PRAM in $O(T(n))$ time.

Definition 3.37
$\text{TP}_{\text{PRAM}}(T(n), P(n))$ is the class of problems solvable by a uniform PRAM with $O(P(n))$ processors in $O(T(n))$ time.

As in the case of the families of uniform circuits, special attention is given to the class $\text{TP}_{\text{PRAM}}(\log^{O(1)} n, n^{O(1)})$.

The following result, relating the power of the PRAM model to the power of deterministic Turing machines, represents one of the many formulations of the thesis of parallel computation (see Section 3.1).

Proposition 3.38 [FW78]
If $S(n) \geq \log n$ then:

$$\bigcup_{k=1}^{\infty} \text{TIME}_{\text{PRAM}}\left(S^k(n)\right) = \bigcup_{k=1}^{\infty} \text{SPACE}_{\text{TM}}\left(S^k(n)\right)$$

Proof. The result proved in [FW78] regarded PRAMs and Turing machines (TMs) as language recognizers. The proof itself consists in simulating a TM with a PRAM, and, conversely, a PRAM with a TM. We will just sketch the first part of the proof, giving the essential information to construct, given a deterministic TM M recognizing a language L in space $S(n) \geq \log n$, a PRAM P accepting L in $O(S(n))$ time. Once more, the fundamental technique is to construct the computation graph of M, and to solve the graph accessibility problem (or, equivalently, to compute the transitive closure of the computation graph of M).

Assume that M has q states and an alphabet made of c symbols. The number of possible configurations is then $N = qnc^{S(n)}S(n)$ (see Proposition 3.23). Let us encode each configuration with a number between 0 and $N_r - 1$, where $N_r = 2^{\lceil \log q \rceil + \lceil \log n \rceil + \lceil S(n) \log c \rceil + \lceil S(n) \rceil}$: the first $\lceil \log q \rceil$ bits encode the state, bits from $\lceil \log q \rceil + 1$ to $\lceil \log q \rceil + \lceil \log n \rceil$ encode the position of the head on the input tape, and so on. It is easy to realize that the encoding and decoding of a TM configuration can be computed in a time proportional to the length of the encoding itself, and therefore in $O(S(n))$ time (being $S(n) \geq \log n$). The computation of the PRAM P uses N_r processors and proceeds as follows.

1. P_i decodes its own identification number, which we assume to be i itself, reconstructing the corresponding configuration in its own local memory, $i = 0, ..., N_r - 1$. For instance, P_i uses register r_0 to store the control state, register r_1 to store head position on the reading tape, etc.
2. P_i simulates a step of the TM starting from the just decoded configuration (see Proposition 2.50), $i = 0, ..., N_r - 1$.
3. If configuration i does not have a successor,[9] P_i sets $j = -1$.
4. If i is a final configuration, P_i sets $j = i$; otherwise it encodes the new configuration value with an integer number $j \in [0, N_r - 1]$ representing the configuration following the configuration i, $i = 0, ..., N_r - 1$.
5. P_i stores the value j in the global memory register m_i, $i = 0, ..., N_r - 1$.
6. At this point, if the input belongs to the language, there is a list of registers, starting with m_0 and ending with register m_f, representing a final configuration (that is, such that the content of m_f is f).
7. If the content of m_i is different from -1, P_i determines the successor in the list, that is, it executes the instructions:

    LOAD @m_i

---

[9]For brevity we say "configuration i" rather than "configuration the encoding of which is i".

STORE r3
           LOAD @r3
           STORE mi

   $i = 0, ..., N_r - 1$. This operation is called *pointer jumping* (see Section 4.3.2).
8. Since the list is at most of length N_r, by iterating step 7 $O(\log N_r)$ times the processor is finally able to reach the end of the list. In this case it accepts the input. Otherwise, if at some point it finds in a register the value -1, it rejects the input.

According to the above discussion on the encoding and decoding time being $N_r = O(S(n))$, the time cost of the simulation is dominated by the same term $O(S(n))$.

As already observed regarding Proposition 3.23, the previous proposition can be extended to the case of PRAMs (and TMs) computing functions and solving problems. To this end, given a function $f:\{0,1\}^n \to \{0,1\}^m$, consider again the language A_f:

$A_f = \{(x, i) \mid \text{the } i\text{-th bit of } f(x) \text{ is } 1\}$

Using a TM M with space cost $S(n) \geq \log n$ computing the function f, one can first construct a machine M' accepting A_f which uses $O(S(n))$ cells of working tape. A PRAM P computing f in $O(S(n))$ time can then be constructed in the following way. P uses mN_r processors, organized in m groups of N_r processors each. On an input x, processors belonging to group i simulate, in $O(S(n))$ time, the machine recognizing A_f on input $<x, i>$. The first processor of group i then writes the result (1 accepted, 0 rejected) to register m$_i$. The value $f(x)$ is then found stored in the first m global memory registers.

An important consequence of Proposition 3.38 is the following.

Corollary 3.39
PSPACE = TIME$_{\text{PRAM}}(n^{O(1)})$, that is, the class of problems solvable sequentially in polynomial space is coincident with the class of problems solvable by a PRAM in polynomial time.

The class PSPACE (see Section 2.5) contains many interesting problems, for instance, all problems in *NP*, for which the best available sequential algorithms have an exponential time cost. The result expressed by Corollary 3.39 could induce the thought that parallelism is a panacea for most computation problems. Note however that the number of processors used in the simulation (Proposition 3.38) is, when $S(n) = n^{O(1)}$, exponential in n.

Such a number of processors is as unacceptable as an exponential time. The real problem seems to be that of a simultaneously restricted request of time and space (number of processors) resource.

Corollary 3.40
The class of problems solvable in logarithmic space by a TM is coincident with the class of problems solvable in logarithmic time by a PRAM with a polynomial number of processors. That is:

$$\text{SPACE}_{\text{TM}}(\log n) = \text{TP}_{\text{PRAM}}(O(\log n), n^{O(1)})$$

Restricting ourselves to the class $\text{SPACE}_{\text{TM}}(\log n)$, still containing many interesting problems, we obtain more relevant indications. The number of processors required by a PRAM to solve problems in this class is in fact at most polynomial in n (that is, exponential in $\log n$). We have therefore, for $\text{SPACE}_{\text{TM}}(\log n)$ in parallel, a restricted resource requirement. Note however that this class of problems behaves well also in the sequential case. In fact, we know that a TM using logarithmic space uses at most polynomial time. For the class $\text{SPACE}_{\text{TM}}(\log^{O(1)} n)$, that is, the class of problems solvable by a TM in polylogarithmic space, things are already worse. Considering only the simulation in Proposition 3.38, we must in fact predict a number of processors $P(n) = O(2^{\log^k n})$, with $k \geq 1$. Therefore, when $k > 1$, the upper bound on the value of $P(n)$ increases faster than any polynomial in n.

3.4.2 Variations on the original model

The examples and the results presented in the previous section should have provided sufficient indications on the architecture of the original PRAM model, and on its computing power. In the literature on parallel algorithms however PRAM programs are almost never as detailed as in Examples 3.1 and 3.3. In most cases a description at a higher level is sufficient and more readable. The following considerations lead to redefining the model in such a way as to "handle" it more easily, above all from the point of view of algorithm design.

1. Initializing a computation with $P(n) = n^{O(1)}$ processors, where n is the input size, requires time $\Theta(\log n)$, as shown in Example 3.31. It is then impossible to have sublogarithmic parallel times for these computations. However, if the computation itself (assuming the processors to be initially active) costs $\Omega(\log n)$ it is then evident that the initialization phase does not influence the order of magnitude of the time cost. This was the case for Example 3.33. Once these points have been cleared, it is absolutely useless to include an initialization phase in every PRAM program. In the following we

will assume that all the $P(n)$ processors are initially active. As a consequence, the FORK instruction is excluded from the instruction set. Also the presence of a status indicator register is not necessary any more.

2. In Example 3.31, the generic processor P_i received, when initialized, a unique identification number j_i, not necessarily coincident with i. It is evident that generality is not lost if we always set $j_i = i$. This hypothesis was used in Example 3.33 and in proving Proposition 3.38. In the following we will assume that each processor can read its own identification number from a special read-only register called PIN (standing for processor identification number).

3. One can assume that the input size n is known to all processors participating in the computation.

4. The most important aspect concerns the (logical) synchronization of processor activity. Even when treating simple examples, such as the computation of $\lceil \log n \rceil$ and the addition of n numbers, it is not easily ensured that all processors interact in a correct way and, above all, there is no guarantee that global memory accesses do not violate some constraint (simultaneous writing for instance). Actually, although the PRAM is synchronous, the programming characteristics are asynchronous (see Chapter 5). For this reason, the original PRAM model has evolved into a model where all processors are forced to execute the same instruction. In other words, there is a single global program counter. Programs are specified using high-level programming languages in which synchronization can be forced in an implicit way. For instance, the instruction:

if PIN = #i **then** $m_i \leftarrow m_i + 1$

executed by *all* processors, is such that only processor P_i assigns $m_i \leftarrow m_i + 1$, whereas all other processors execute a synchronization loop. The same is true for iterative constructs.

In the following we will discuss two other architectural aspects considered in the literature, providing a series of possible variations on the original model.

Instruction set

As in the RAM case (see Section 2.4.4), the thesis of sequential computation gives some conditions for judging the suitability of a given instruction set.

1. Each PRAM processor must have a computing power not greater than that of a reasonable sequential machine. As a consequence, according to the thesis of sequential computation, each instruction in the set of a parallel machine operating within the time bound $S(n)$ must be amenable to simulation by a deterministic TM in at most $S(n)^{O(1)}$ steps.

In the PRAM case, however, we must also take into account the thesis of parallel computation.

2. Each instruction of a parallel machine operating within the time bound $S(n)$ must be amenable to simulation by a deterministic TM using $S(n)^{O(1)}$ working tape cells.

The instruction set of the PRAM presented in Section 3.4.1 satisfies the above two conditions, even when the uniform cost criterion is adopted, if $S(n) \geq \log n$. The time needed by a Turing machine to simulate addition (or subtraction) is in fact $O(\max(\log n, S(n)))$, the same bound holding for space. The following extensions have in any case been considered in the literature.

1. *Minimal* instruction set. With respect to the classic PRAM set, the division by 2 of a register content is also included (corresponding to shifting binary numbers one position to the right):

    shift ri $r_i \leftarrow \lfloor r_i/2 \rfloor$

    Conditions 1 and 2 above are evidently valid also for the minimal repertory.

2. *Restricted* instruction set. This includes, beside those of the minimal set, an arbitrary shift instruction, either positive or negative (corresponding to multiplication by a power of 2):

    shift ri $A \leftarrow \lfloor A * 2^{r_i} \rfloor$

    With the restricted instruction set, conditions 1 and 2 are valid only when a cost is ascribed to every instruction including the length of operand representation. Alternatively, to use the unit cost measure, it is necessary to impose a bound of the type $W(n) = n^{O(1)}$, where $W(n)$ is the maximum number of bits in the representation of the operands.

3. *Full* instruction set. It corresponds to the restricted set with the addition of multiplication and division:

   mul ri $A \leftarrow A * r_i$
   div ri $A \leftarrow \lfloor A/r_i \rfloor$

The considerations relative to the restricted instruction set also apply to this case.

Example 3.41 (computation of $\lfloor \log n \rfloor$ and $\lceil \log n \rceil$)
Using the restricted instruction set and the unit cost measure, a PRAM with at least $P(n) = \lceil \log n \rceil$ processors can compute $\lfloor \log n \rfloor$ and $\lceil \log n \rceil$ in constant time. Each processor executes the following algorithm, where the condition expressed by the outermost **if** is satisfied if and only if PIN $= \lfloor \log n \rfloor$. The result can be read from registers m_0 and m_1.

begin
     $r_0 \leftarrow \lfloor n/2^{PIN} \rfloor$;
     if $r_0 > 0$ **and** $\lfloor r_0/2 \rfloor = 0$ **then begin**
         $m_0 \leftarrow$ PIN;
         if $n = 2^{PIN}$ **then** $m_1 \leftarrow$ PIN **else** $m_1 \leftarrow$ PIN $+ 1$
     end
end

A deterministic TM cannot simulate an arbitrary shift instruction in a constant space. In fact, if n is the shift width (number of positions by which the content of the storage register must be shifted), the machine must evidently be able to count up to n, which entails the use of at least $\lceil \log n \rceil$ cells of working tape. Therefore, adopting the restricted instruction set does not satisfy condition 2.

Global memory access

In the original PRAM more than one processor can simultaneously read the contents of the same global memory register, while every attempt at simultaneous writing is stopped immediately by halting the computation. A PRAM following this criterion to discipline global memory access is called *CREW PRAM*, where CREW stands for *Concurrent Read Exclusive Write*. Two other solutions, both present in the literature, are possible. In the *EREW PRAM* read attempts from the same register are also stopped. Conversely, in the *CRCW PRAM* model simultaneous writing to the same memory location is allowed. In this case, to ensure that the computation is determined, a criterion is required to establish uniquely what the content of the global memory cell will be at the end of the write operations. The following CRCW PRAM versions appeared in the literature.

- *Weak.* More than one processor can write simultaneously to the same location only if the value written is 0.
- *Common mode.* More than one processor can write simultaneously to the same location only if each one presents the same value.
- *Arbitrary winner.* The value written is arbitrary, and it cannot be determined *a priori* which of the values presented by the processors it will be.
- *Priority mode.* The value written is the one presented by the processor characterized by the largest (or equivalently the smallest) PIN.
- *Strong.* The value written is the largest (or equivalently the smallest).

Example 3.42 (GAP)

The accessibility problem of a direct graph G with n vertices can be solved in time $T(n) = O(\log n)$ on a common mode CRCW PRAM having a full instruction set and using $P(n) = n^3$ processors. The input to the machine is given in the form of the adjacency matrix A of G. We know (see Section 2.5) that the product matrix $A^k = (a_{ij}^{(k)})$ describes paths in the graph having length at most k. In other words, $a_{ij}^{(k)} = 1$ if a path exists in G, having length at most k, linking nodes i and j, and $a_{ij}^{(k)} = 0$ otherwise. Since a loopless path linking any two nodes has maximum length $n - 1$, the problem of verifying whether s is linked to t can be solved by computing A^k, $k \geq n - 1$, and checking if $a_{st}^{(k)} = 1$. Each of the n^3 processors participating in the computation first gives, according to its own PIN, a unique value to the indices i, j and h ($0 \leq i, j, h \leq n - 1$). After this operation, the n^3 processors can be identified using the three index notation P_{ijh} ($0 \leq i, j, h \leq n - 1$). The computation itself is a loop of $\lceil \log n \rceil$ steps where, at step k, processor P_{ijh} computes the logic product $a_{ih}a_{hj}$ and, if the result is true (= 1), sets $a_{ij} = 1$, meaning that there exists, between nodes i and j, a path G of maximum length 2^k. The following program computes the matrix $A^{2^{\lceil \log n \rceil}}$ solving the accessibility problem for any pair of graph nodes (see [Pa87]). For the sake of simplicity, the program uses symbolic names in place of registers.

```
begin
    i ← ⌊PIN/n²⌋;
    j ← ⌊(PIN – i * n²)/n⌋;          comment j ← ⌊(PIN mod n²)/n;
    h ← PIN – ⌊PIN/n⌋ * n;            comment h ← PIN mod n;
    k ← 1;
    while k < n do begin
        if a_ih * a_hj = 1 then a_ij ← 1;
```

           $k \leftarrow k * 2$
       end
   end

Note that the only simultaneous writing to global memory locations is $a_{ij} \leftarrow 1$. Since all processors write the same value, the program is in fact of common mode type.

By carefully considering Example 3.42 one can realize that the classic PRAM instruction set is sufficient to solve the problem in $O(\log n)$ time. Note indeed that all values used in the computation can be represented by $O(\log n)$ bits. From the results given at the end of Section 3.3, we know that addition and multiplication of log n-bit numbers can be implemented by families of uniform Boolean circuits of depth $O(\log\log n)$, while division of log n-bit numbers can be implemented by families of uniform Boolean circuits of depth $O((\log\log n)^2)$. Since even the weakest of the PRAM models (the EREW) is at least as powerful as the uniform circuit model (see Section 3.4.3), the first three steps of the program in Example 3.42 can be done in $O((\log\log n)^2)$ time by a PRAM with a classic instruction set (that is, addition and subtraction only).

The accessibility problem of a graph can be solved by a TM in polynomial time and $O(\log^2 n)$ space (see Section 2.5). We have then a problem which, in parallel, is amenable to a solution exponentially faster than in the sequential case.

Example 3.43 (computing the maximum among n numbers)
A weak CRCW PRAM with at least n^2 processors and a full instruction set can compute the maximum among n numbers, $x_0, ..., x_{n-1}$, in constant time. To this end, it is sufficient to subdivide the n^2 processors into n groups of n processors each. Initially, the first processor of each group (that is, processors P_{i*n}, $i = 0, ..., n - 1$) writes the value 1 to a preset global memory location, be it f_i, $i = 0, ..., n - 1$. Following this, the jth processor of the ith group ($i, j = 0, ..., n - 1$) computes the difference $d_{ij} = x_i - x_j$. If $d_{ij} < 0$, or if $d_{ij} = 0$ and $i < j$, it then signals this fact by writing the value 0 to the location f_i. It is obvious that if x_k is the greatest among the n numbers (and there is no index $h > k$ such that $x_h = x_k$), at the end of this step location f_k will still contain the value 1, whereas locations f_i, $i = 0, ..., k - 1$, $k + 1, ..., n - 1$, will contain the value 0. At this point the first processor in each group reads the corresponding location f_i, and the processor in group k gives the result. The algorithm is as follows.

   begin
       $i \leftarrow \lfloor PIN/n \rfloor$;   **comment** computing the group number;

```
        j ← PIN - i * n;           comment computing the number in the group;
        if j = 0 then f_i ← 1;
        d_ij = x_i - x_j;
        if d_ij < 0 or (d_ij = 0 and i < j) then f_i ← 0;
        if j = 0 then begin         comment processor P_{n*i} checks if f_i = 0;
           if f_i = 0 then result ← x_i
        end
     end
```

The previous example calls attention to the important problem of determining non-trivial lower complexity bounds for models such as the CRCW PRAM. For other computation models (the CREW PRAM itself, and, above all, Boolean and arithmetic circuits), it is in some cases possible to determine a lower complexity bound by means of simple considerations on operation fan-in (see Section 3.3). For instance, it is obvious that iterated addition (that is, addition with n numbers) is dependent on every number to be added. Considering elementary binary operations one can easily verify that the minimum parallel time needed only to "combine" the n numbers in a single value is $\lceil \log n \rceil$, following the scheme of the algorithm in Example 3.33, which turns out to be asymptotically optimal. Algorithms, as in the said example, computing functions of the type $f(x_1, ..., x_n) = x_1 @ x_2 @ ... @ x_n$, where @ is any associative operation, are called *fan-in* algorithms. In the case of a CRCW PRAM, arguments based on operation fan-in are useless. Consider the algorithm in Example 3.43. It can be easily seen that the AND of the n comparisons $x_i < x_j, j = 0, ..., n - 1$, computed by the processors in the ith group, is in fact written to each memory cell f_i ($i = 0, ..., n - 1$) in constant time. This means that, since n can increase indefinitely, in a CRCW PRAM one can compute logic operations with unbounded fan-in [SV84]. Non-constant lower complexity bounds have been, on the other hand, determined for a CRCW PRAM in the case of limited global memory [VW83].

3.4.3 Relations among PRAM models

In this section we present some results relating the power of the various PRAM models introduced in Section 3.4.2, from EREW to strong CRCW. We will also briefly discuss the relations linking PRAMs to uniform circuits.

We must first extend Definitions 3.36 and 3.37.

Definition 3.44
$\text{TIME}_{\text{EREW}}(T(n))$ ($\text{TIME}_{\text{CREW}}(T(n))$) is the class of problems solvable by a uniform EREW (resp. CREW) PRAM in $O(T(n))$ time.

Definition 3.45
$TP_{EREW}(T(n), P(n))$ ($TP_{CREW}(T(n), P(n))$) is the class of problems solvable by a uniform EREW (resp. CREW) PRAM with $O(P(n))$ processors in $O(T(n))$ time.

For CRCW PRAMs we must also specify the criteria used to solve simultaneous write conflicts.

Definition 3.46
$TIME_{weak}(T(n))$ (resp. $TIME_{com}(T(n))$, $TIME_{arb}(T(n))$, $TIME_{prior}(T(n))$, $TIME_{str}(T(n))$) is the class of problems solvable by a weak (resp. common mode, arbitrary winner, priority mode and strong) CRCW PRAM in $O(T(n))$ time.

Definition 3.47
$TP_{weak}(T(n), P(n))$ (resp. $TP_{com}(T(n), P(n))$, $TP_{arb}(T(n), P(n))$, $TP_{prior}(T(n), P(n))$, $TP_{str}(T(n), P(n))$) is the class of problems solvable by a weak (resp. common mode, arbitrary winner, priority mode and strong) CRCW PRAM in $O(T(n))$ time using $O(P(n))$ processors.

Proposition 3.48
For any $T(n)$ the following relations hold:

1. $TIME_{EREW}(T(n)) \subseteq TIME_{CREW}(T(n))$
2. $TIME_{CREW}(T(n)) \subseteq TIME_{weak}(T(n))$
3. $TIME_{weak}(T(n)) \subseteq TIME_{com}(T(n))$
4. $TIME_{com}(T(n)) \subseteq TIME_{arb}(T(n))$
5. $TIME_{arb}(T(n)) \subseteq TIME_{prior}(T(n))$
6. $TIME_{prior}(T(n)) \subseteq TIME_{str}(T(n))$

Proof. Relations 1–5 are obvious. Indeed, in each relation the model on the left-hand side is a particular case of the model on the right-hand side. Thus, for instance, an EREW PRAM program is a particular case of a CREW PRAM program. With respect to relation 6 we note that a strong CRCW PRAM can simulate every step of a priority mode CRCW PRAM in constant time. The problem is obviously limited to simulating a step of simultaneous write. The goal is to ensure that the value written is that coming from the processor with the greatest PIN. To this end, all processors needing to write to a given cell, say m_i, write first their own PIN to a second established cell, m_{i+k} for instance, with a suitable k. The value written to m_{i+k} identifies the processor which, in a priority mode CRCW, would have prevailed in the write conflict on m_i. Each processor must then compare the contents of m_{i+k} with its own identifier, and, if the two values coincide, it must write the original value to m_i.

The above proposition, which holds even in the case of a simultaneous bound on the time and number of processors resources, proves that the seven models we have discussed in the present chapter constitute a hierarchy. We now turn to the inverse problem, that is, the precise determination of the gain in computing power when going from one model to the next in the hierarchy. The result expressed by Proposition 3.51 is of fundamental importance. In fact, relating the EREW to the strong CRCW model (that is, the last and the first model in the hierarchy) gives global indications also on intermediate models. Proposition 3.51 uses results given in Propositions 3.49 and 3.50.

Proposition 3.49 [Co86]
The sorting of n numbers can be implemented on an EREW PRAM in O(log n) parallel time using O(n) processors.

Proof. See Section 4.3.6.

Proposition 3.50
Let @ be any associative operation defined on the elements of a set X, and let $<x_1, x_2, ..., x_n>$ be a sequence of n elements of X. For any $k \in [1,n]$, denote with S_k the quantity x_1 @ x_2 @ ... @x_k. S_k is called *the kth prefix sum* of $<x_1, x_2, ..., x_n>$. The computation of S_k, $k = 1, ..., n$, can be done in O(log n) time on an EREW PRAM using O(n/log n) processors.

Proof. See Section 4.3.1.

While proving Proposition 3.51 we will need to compute the prefix sum of a sequence where: (1) each element is a pair of numbers, (2) the associative operation with respect to which the sums are computed is the lexicographic maximum among number pairs.[10] For instance, the six prefix sums in the sequence $<(2,3), (1,5), (3,3), (1,8), (3,5), (2,7)>$ are (2,3), (2,3), (3,3), (3,3), (3,5), (3,5).

Proposition 3.51 [NS81]
A parallel computation which can be performed on a strong CRCW PRAM in time $T(n)$ using $P(n)$ processors can also be performed on an EREW PRAM in O($T(n)$log $P(n)$) time using O($P(n)$) processors.

Proof. Let us describe how a write or read step of the strong CRCW PRAM can be simulated by an EREW PRAM. An array A consisting of $P(n)$ elements is used in the simulation. Each of these elements is a triplet of numbers (processor number, memory address, value to be read/written). For

---

[10]Recall that the pair of numbers (a, b) is lexicographically greater that the pair (c, d) if and only if $a > c$ or if $a = c$ and $b > d$.

the sake of simplicity, we will denote by $A_i(k)$, $A_a(k)$ and $A_v(k)$ the three elements of the triplet in the position $A(k)$, $k = 0, ..., P(n) - 1$. Each processor can access this structure using its own PIN. To simulate a read (resp. write) step, processor P_k, $k = 0, ..., P(n) - 1$, writes to $A_i(k)$ the value k and to $A_a(k)$ the address where it has to read from (resp. write to). Furthermore, if the step to be simulated is a write step, P_k stores in $A_v(k)$ the value to be written (otherwise $A(k)$ remains unspecified for the moment).

At this point, triplets are sorted with respect to their address. In the case of a write step, triplets with the same address are sorted with respect to their value (from least to greatest). According to Proposition 3.49, the sorting can be carried out in $O(\log P(n))$ time using $O(P(n))$ processors.

Once the array A has been sorted, the write operation can be simulated in the following way. Processor P_k compares $A_a(k)$ to $A_a(k + 1)$ (accessing position $A(k)$ first and then position $A(k + 1)$ to avoid conflicts), $k = 0, ..., P(n) - 2$. If the two addresses are different, P_k writes the value $A_v(k)$ to the register with address $A_a(k)$. This operation is always executed by processor $P_{P(n)-1}$.

Simulating a read step is somewhat more complex. Processor P_k compares $A_a(k)$ to $A_a(k - 1)$ (accessing position $A(k)$ first and then position $A(k - 1)$), $k = 1, ..., P(n) - 1$. If the two addresses are different P_k reads the value contained in the register with address $A_a(k)$, which then becomes the third element, $A_v(k)$, of the triplet. This operation is always executed by P_0. When $A_a(k) = A_a(k - 1)$, processor P_k sets as the third element of the triplet the lowest number which can be contained in a machine register, $k = 1, ..., P(n) - 1$. With this operation, all the values to be read are found in well-determined elements of array A. The successive operation consists of distributing the values in the remaining elements, in such a way that equal values correspond to equal addresses. To this end, the processors compute, in parallel $O(\log P(n))$ time, the lexicographic maximum among all pairs $(A_a(j), A_v(j))$ contained in each prefix of length k (that is, in the first k positions) of the array, $j = 0, ..., k - 1$, $k = 1, ..., P(n)$. The situation is shown in Figure 3.17, in the special case $P(n) = 8$. Part (a) of the figure shows the array A, while part (b) shows the array B obtained from the computation of the lexicographic maximum on the prefixes of A (note that $B_i(k) = A_i(k)$). The symbol "−" denotes the minimum value which can be represented in a memory register. The simulation is completed as follows. Processor P_k writes the triplet $B(k)$ to the position $B_i(k)$ of an auxiliary array C and then reads the value $C_v(k)$, this being the desired one, $k = 0, ..., P(n) - 1$.

An immediate consequence of Proposition 3.51 is the fact that any computation which can be carried out in polylogarithmic time with a polynomial number of processors on any one PRAM model can also be carried out in polylogarithmic time with a polynomial number of processors on all PRAM models. More precisely, the following result holds.

I	A	V
3	6	-1
2	8	5
6	8	-
7	8	-
1	10	12
5	11	-5
0	18	2
4	18	-

(a)

I	A	V
3	6	-1
2	8	5
6	8	5
7	8	5
1	10	12
5	11	-5
0	18	2
4	18	2

(b)

Figure 3.17 Computation of the lexicographic maximum on array prefixes.

Corollary 3.52
For any $k \geq 1$ it is:

$$TP_{EREW}(\log^k n, n^{O(1)}) \subseteq TP_{str}(\log^k n, n^{O(1)}) \subseteq TP_{EREW}(\log^{k+1} n, n^{O(1)})$$

We will end the discussion of the PRAM model by mentioning the relations among uniform PRAMs and families of uniform circuits. It can be proven that, for any $k \geq 1$, the following chain inclusion holds:

$$NC^k \subseteq TP_{EREW}(\log^k n, n^{O(1)}) \subseteq TP_{prior}(\log^k n, n^{O(1)}) \subseteq NC^{k+1}$$

A direct corollary is that the class NC is coincident with the class of problems solvable in polylogarithmic time on any PRAM model using a polynomial amount of processors. This is evidently a strong reason to justify the importance of the class NC. The foregoing relation can also be used to establish correspondences among complexity classes relative to the PRAM model and sequential complexity classes.

3.5 Structured models

In a *structured* model the objects to be manipulated, input data, final or intermediate results, belong to a well-defined mathematical domain. The structured models are distinct from *general* models, as Boolean circuits or Turing machines, which can be seen as string processors on a given alphabet.

The difference between structured and general models, which is discussed in depth in [Bo82], is a relevant one. In a structured model the problem of the representation is not the most important one. In particular, it is not necessary that the elements of the domain can be encoded as finite-length strings on a given alphabet. It could appear that these models do not enable one to draw general conclusions on complexity. The fundamental arguments, as the problem $P = ?NP$ or the result $\text{SPACE}_{TM}(\log^{O(1)} n) = \text{DEPTH}(\log^{O(1)} n)$, are actually pertinent to general models. Structured computation environments are however interesting for at least three reasons (see [Bo82]):

1. Many problems can be described in a natural way using a structured model. The results obtained can be related to the general theory by a detailed study on the representation of the domain elements.
2. By studying structured models, techniques can be developed which are also interesting in the case of general models, proof techniques for instance.
3. Finally, the possibility always exists that a structured model could give general results on well-defined complexity issues. An example is given by the relation between the computation of the nth power of a matrix of order n, with elements on a generic ring, and the transitive closure of a graph. It is well known that the computation of the nth power can be implemented in $O(\log^2 n)$ time (see Section 3.5.1). A result improving this limit would affect the computation of the transitive closure and, therefore, would improve the result in [Sa70] regarding the space sufficient to simulate non-deterministic computations[11] (see Section 2.5).

We will consider below two important structured models: arithmetic circuits and parallel comparison trees.

3.5.1 Arithmetic circuits

Many concepts relative to Boolean circuits also apply to arithmetic circuit theory. An arithmetic circuit α on a field F is a directed, labelled and acyclic graph. Its nodes are subdivided into input, constant, operation and output nodes. The constant nodes are labelled by elements of the field F, whereas each operation node is labelled by one of the four symbols $\{+, -, *, /\}$, denoting the basic operations defined on F. An arithmetic circuit computes a function in a way similar to a Boolean circuit. In this case however the function will have domain $F^{g(n)}$ and codomain $F^{h(n)}$, where n is the problem size. In other words, the computation takes place within the mathematical structure defined by F.

---

[11]It is however necessary that the computation of the power be executable without generating exponentially increasing numbers. In general it is possible that the elements of the nth power of a matrix with only 0s and 1s are larger than of 2^n (see [Bo77]).

108 Parallel computation models

Definition 3.53
A family of arithmetic circuits is a collection of arithmetic circuits $\{\alpha_n\}_{n \in N}$, where each α_n has $g(n)$ inputs and $h(n)$ outputs. If, for any $n \in N$, circuit α_n computes the function $f_n: F^{g(n)} \to F^{h(n)}$, we will say that $\{\alpha_n\}_{n \in N}$ computes $f = \langle f_n \rangle_{n \in N}$.

Any single function f_n can be given as a collection of $h(n)$ unary functions $f_n^{(1)}, ..., f_n^{(h(n))}$. Since the only primitive operations are the four arithmetic operations, each $f_n^{(i)}$ is a *rational function*. We are here interested, contrary to the Boolean case, in the computation of functions rather than in the solution of problems: in the arithmetic context many problems have a single solution. There are however some algebraic/combinatorial problems which are amenable to more than one solution, for instance, the problem of determining a subset of linearly independent columns of a matrix.

The size and depth concepts can be trivially extended to arithmetic circuit families.

Example 3.54 (sum of n numbers)
The circuit shown in Figure 3.18 represents a family of circuits which computes the sum of $n = 2^k$ numbers on a field F. If n is not a power of 2, zeros must be added until the first power of 2 greater than n is reached ($2^{\lceil \log n \rceil} - n$ to be exact). It is easily seen that the size and depth of the family are, neglecting possible additive constants, $n + 2^{\lceil \log n \rceil}$ and $\lceil \log n \rceil + 2$ respectively.

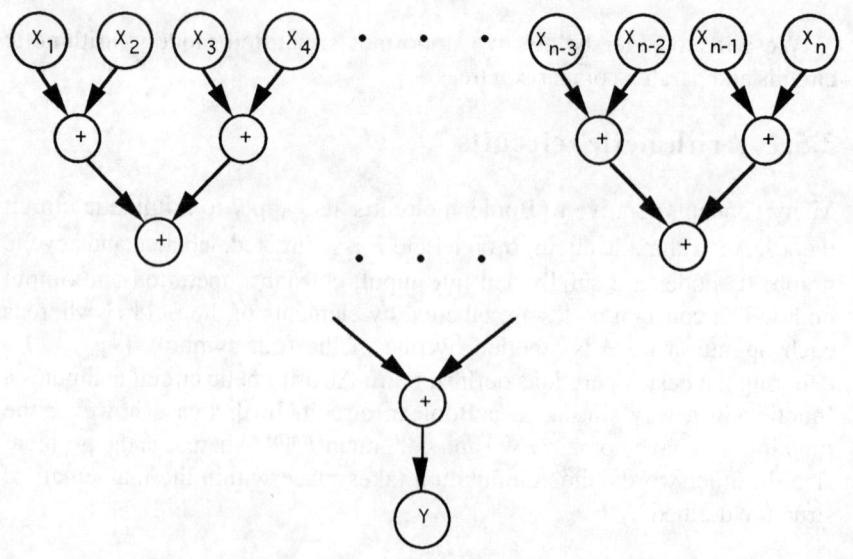

Figure 3.18 Family computing the sum of n numbers.

Structured models 109

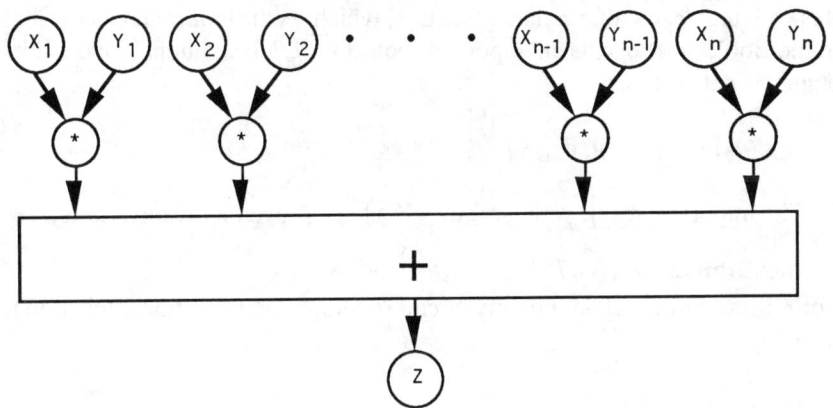

[Figure: arithmetic circuit diagram with inputs $X_1, Y_1, X_2, Y_2, \ldots, X_{n-1}, Y_{n-1}, X_n, Y_n$ feeding pairs of multiplication nodes, all feeding a summation block with output Z]

Figure 3.19 Family computing the scalar product of two n-dimensional vectors.

Example 3.55
Using the circuits from the previous example, one can implement a family of circuits computing the scalar product z between two n-dimensional vectors $\mathbf{x}$ and $\mathbf{y}$. Figure 3.19 shows the generic circuit of the family, which is characterized by size $3n + 2^{\lceil \log n \rceil}$ (neglecting possible additive constants) and depth $\lceil \log n \rceil + 3$.

Example 3.56
The product Z of two matrices X and Y of order n, with elements in F, can be computed by an arithmetic circuit on F of depth $O(\log n)$ using n^2 copies of the circuit in Figure 3.19. The element z_{ij} is computed by the $(n(i-1) + j)$th pair, as the scalar product between the ith row of X and the jth column of Y. A part of the resulting circuit is shown in Figure 3.20, where the block labelled $\langle X_{i*}, Y_{*j} \rangle$ is the operation part of the circuit in Figure 3.19. Size and depth of the circuit are $O(n^3)$ and $O(\log n)$ respectively.

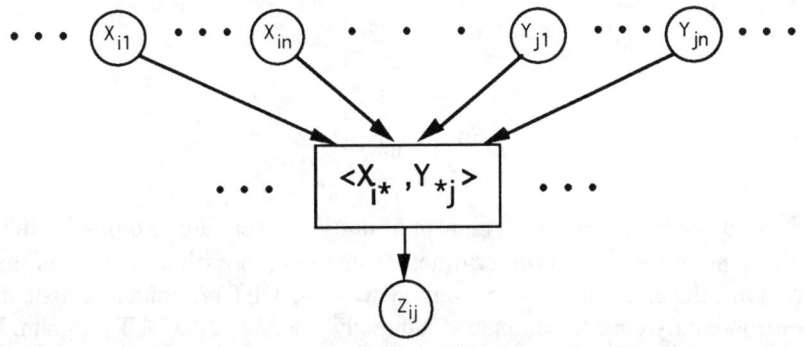

[Figure: inputs $X_{i1}, \ldots, X_{in}, Y_{j1}, \ldots, Y_{jn}$ feeding a block labelled $\langle X_{i*}, Y_{*j} \rangle$ with output z_{ij}]

Figure 3.20 Family of arithmetic circuits computing the product of two $n \times n$ matrices.

Arithmetic circuits can be used both in algorithm design and in the proof of lower bounds on the computational complexity of rational functions. To this end, the *degree* of a rational function, which seems to have no equivalent in the Boolean theory, is an important tool. If $f = g/h$ is a rational function its degree is defined as:

$$\deg f = \max\{\deg g, \deg h\}$$

The important result which follows has been proven by Kung [Ku76].

Proposition 3.57 [Ku76]
Let F be an infinite field. For any circuit α computing the rational function f, it is:

$$D(\alpha) \geq \log \deg f$$

The above result shows that the degree plays a fundamental role in the computational complexity of rational functions.

The uniformity problem for arithmetic circuits needs a detailed analysis. Recall that a family is said to be uniform when there exists a reasonably efficient algorithm able to generate a description of the single circuits. In the case of arithmetic circuits, besides the description of the structure, one also has the problem of describing the elements of the field F to be used as constants. In fact, according to the nature of F, some constants used in the circuit could lack a finite explicit representation. In other cases a large use of computational resources (of the generator) may be necessary to determine the constants. As an example, consider the DFT (discrete Fourier transform) of an n-dimensional vector on the field of complex numbers. A program to compute the DFT will use the following (Vandermonde) matrix:

$$V = \begin{bmatrix} 1 & 1 & 1 & 1 & \ldots & 1 \\ 1 & \omega & \omega^2 & \omega^3 & \ldots & \omega^n \\ 1 & \omega^2 & \omega^4 & \omega^6 & \ldots & \omega^{2n} \\ \ldots & \ldots & \ldots & \ldots & \ldots & \ldots \\ 1 & \omega^n & \omega^{2n} & \omega^{3n} & \ldots & \omega^{n^2} \end{bmatrix}$$

where ω is a primitive $(n+1)$th root of unity.[12] There are two possibilities: either the programmer determines (somehow, possibly using another program) the elements of V, passing them to the DFT program as constants, or the auxiliary code to compute V is directly inserted in the DFT program. In both cases the solution has a non-vanishing cost. The same principle must be applicable to arithmetic circuits.

---

[12] A nth root of unity on the complex field is a complex number ω such that $\omega^n = 1$.

Regarding the representation, note that it is assumed that an arithmetic circuit operates with the exact elements of F, and not with finite approximations. It is clear that when F is finite, or it is the field of rational numbers, its elements can be represented by finite-length strings on a well-defined alphabet. If this is not the case, implicit representations of the constants can be assumed, for instance as solutions of algebraic equations [Ga86]. The nth roots of unity then could be represented as solutions of the equation $X^n - 1 = 0$. In this way only a few constants (namely 0 and 1) would have to be represented explicitly.

Example 3.58. [Ga86]
Using only the constant 2, it is possible, by means of repeated multiplications, to obtain large constants: $2, 4, 16, ..., 2^{2^n}$. A circuit using such constants could have as a component the circuit shown in Figure 3.21.

The circuit computing $\{2, 4, 16, ..., 2^{2^n}\}$ has size and depth $2^{n+1} - 1$ and n respectively.

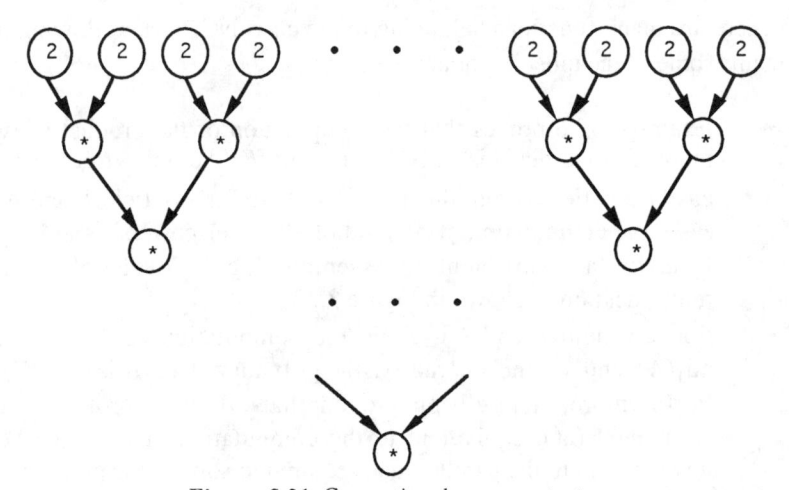

Figure 3.21 Computing large constants.

Following the above arguments we give below the definition of uniformity for families of arithmetic circuits. At least when $F = Q$, the definition appears to be general.

Definition 3.59
A family $\{\alpha_n\}_{n \in N}$ of arithmetic circuits of size $C(n)$ and depth $D(n)$ is log space-uniform (*P*-uniform) if there appear in it explicitly only the constants 0 and 1, and if the structure of α_n can be computed by a deterministic Turing machine using $O(\log(C(n)))$ cells of working tape (resp. $C(n)^{O(1)}$ steps).

As in the Boolean case, the class of functions that can be computed by circuits of polylogarithmic depth and polynomial size in the number of inputs is of special interest. The following definitions introduce the arithmetic analogs of the classes NC^k and NC.

Definition 3.60

NC_F^k (NC_F^k (P-uniform)) is the class of rational functions in $\cup F(X_1, ..., X_n)$ that can be computed by log space-uniform (resp. P-uniform) families of arithmetic circuits of size $n^{O(1)}$ and depth $O(\log^k n)$.

Definition 3.61

$$NC_F = \bigcup_{k \geq 1} NC_F^k$$

and

$$NC_F(P\text{-uniform}) = \bigcup_{k \geq 1} NC_F^k(P\text{-uniform})$$

There are many functions belonging to the class NC_F. The following list contains some of the most important ones.

- Example 3.56 proves that the computation of the product of two matrices of order n is in NC_F^1, for any F (the uniformity can be easily satisfied). From this it follows that, for any field F and $A \in F^{n \times n}$, the computation of A^n (and of all lower powers) is in NC_F^2. It is in fact sufficient to assemble $\log n$ levels of matrix multiplication as shown in Figure 3.22.
- For any field F and $A \in F^{n \times n}$, the computation of $\det(A)$, A^{-1}, $\text{adj}(A)$ and of the characteristic polynomial of A is in NC_F^2. Furthermore, it has been proven that all these problems are equivalent (and equivalent to the computation of A^n). The best lower bound to the parallel time required to solve these problems is $2*\log n$, which follows immediately from arguments on fan-in (see Section 4.4 and [Cs76], [Be84], [BGH82]).
- For any field F, the computation of the product of two polynomials on $F[x]$ is in NC_F^1.

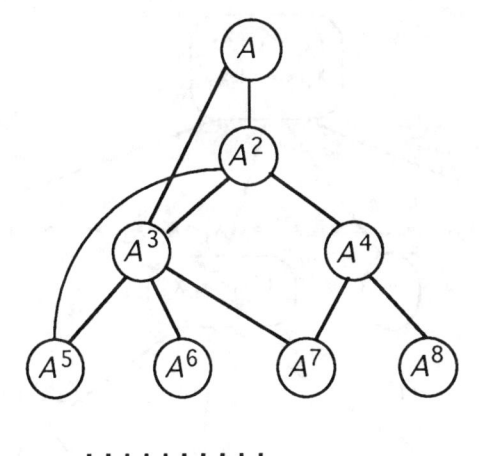

Figure 3.22 Scheme to compute the powers of a matrix.

3.5.2 A model for comparison problems

Let S be a linearly ordered set, and let $\leq$ denote the ordering relation. Given n elements of S, $x_1, ..., x_n$, not necessarily different, the following problems can be expressed in terms of the relation $\leq$, independently of the nature of the elements of S:

- Determine $j \in [1, n]$, such that, for $i = 1, ..., n$, $i \neq j$, it is $x_j \leq x_i$ (or $x_j \geq x_i$).
- Determine a permutation $j_1, j_2, ..., j_n$ of $[1, n]$, such that it is $x_{j_1} \leq x_{j_2} \leq ... \leq x_{j_n}$.

The first one is a minimum (maximum) problem, while the second one is a sorting problem. A computation model oriented towards the solution of these problems will involve only comparison operations to find the relations between pairs of elements of S, and their sequencing.

The model of parallel computation which we will now consider is a generalization of a similar model of sequential computation, the comparison tree (or decision tree, see [AHU74]). A *parallel comparison tree* idealizes a situation where p processors are available to execute as many comparison operations on the elements of S. More precisely, each processor executing a comparison operation on the elements x_i and x_j determines, in constant time, which of the relations $x_i < x_j$, $x_i = x_j$ or $x_i > x_j$ is satisfied. We will denote this type of comparison with $\{<,=,>\}$. At the end of each step the algorithm decides, based on the result, the p pairs to be compared in the next step. The situation can be sketched using a 3^p-ary tree, where each node is labelled by a set of at most p comparisons $\{<,=,>\}$, and with at most 3^p outgoing arcs, each labelled by one of the possible results.

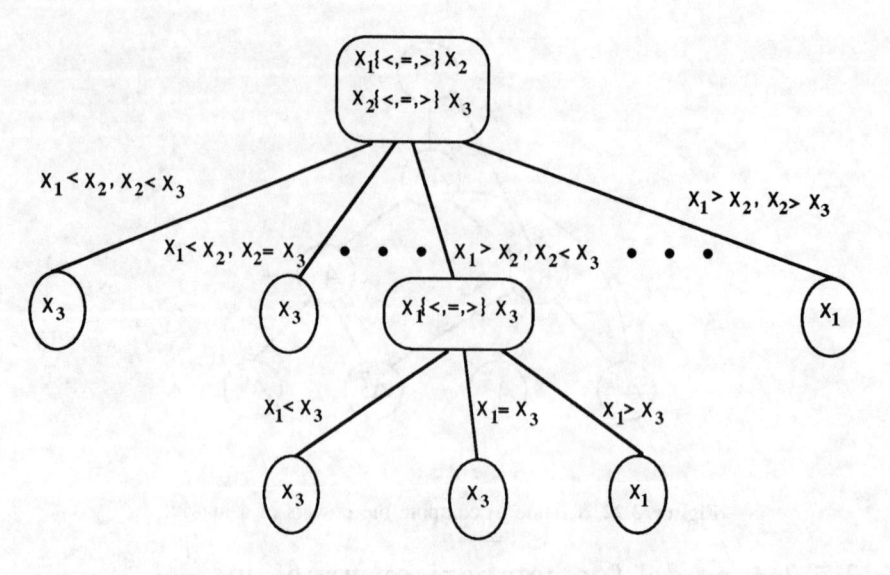

Figure 3.23 Parallel comparison tree for Example 3.28.

Example 3.62
If $p = 2$, the parallel comparison tree in Figure 3.23 represents an algorithm to sort $n = 3$ elements of a linearly ordered set.

Besides the number of processors, which determines the maximum number of comparisons that can be done in parallel, the other important complexity measure for this model is depth. This is the maximum length of a path linking the root with a leaf. The depth of a given tree represents the number of comparisons in the worst case: it is a measure of the time cost of the algorithm described by the tree.

The use of parallel comparison trees is well suited to the determination of lower bounds on the complexity of this type of problem.

Definition 3.63
$MAX_p(n)$ (resp. $ORD_p(n)$) is the minimum depth of all parallel comparison trees with p processors solving the problem of determining the maximum of (resp. sorting) n elements of a linearly ordered set.

The following results have been proved in [Va75].

Proposition 3.64
If $1 < p < n$ it is:

$$MAX_p(n) = \Omega(n/p + \log\log p)$$

If $p = n > 1$ it is:

$MAX_p(n) = \Omega(\log\log n)$

If $4 \leq 2n \leq p \leq n(n-1)/2$, it is:

$MAX_p(n) = \Omega(\log\log n - \log\log(p/n))$

Note that, if $p = n(n-1)/2$, it is possible to execute in a single CRCW PRAM step all possible comparisons, and to obtain the value of the maximum element in constant time (see Example 3.43).

Since the sequential sorting of n elements requires $\Omega(n\log n)$ comparisons (see Example 2.47), with $O(n)$ available processors an immediate lower bound on $ORD_p(n)$ is $\Omega(\log n)$.

Proposition 3.65 [HH81]
If $p = O(n\log^\alpha n)$, $\alpha > 0$, it is:

$ORD_p(n) = \Omega(\log n /\log\log n)$

Optimal algorithms for maximum determination are developed in [Va75]. These algorithms reach the bounds given in Proposition 3.64. However, being an idealized model, the parallel comparison tree neglects some important problems arising on actual models of parallel machines (the PRAM in Section 3.4 for instance). In many cases, the cost of solving these problems cannot be neglected. The main problem we refer to is known in the literature as the *processor allocation* problem. Note that in a parallel comparison tree processors are anonymous. Only the (at most) p comparisons executed at each step are important. In the PRAM case, there are also the problems of exactly specifying data accessed by every single processor at each step, and of communicating the results. From this point of view, the result by Valiant has been subsequently adapted by various authors to the PRAM model.

Proposition 3.66 [SV81]
On a common mode CRCW PRAM:

if $p \leq n$ it is:

$MAX_p(n) = O(n/p + \log\log p)$

if $n \leq pn(n-1)/2$ it is:

$MAX_p(n) = O(\log\log n - \log\log(p/n + 1))$

Regarding the sorting problem, the two results below are optimal with respect to the product (time) x (number of processors). Optimality is an immediate consequence of the lower bound $\Omega(n\log n)$ of the same problem in the sequential case.

Proposition 3.67 [BN86]
On an EREW PRAM, if $p = n/\log n$ it is:

$ORD_p(n) = O(\log^2 n)$

Proposition 3.68 [Co86]
On an EREW PRAM, if $p = n$ it is:

$ORD_p(n) = O(\log n)$

It is interesting to note that the first optimal (in the sense stated above) sorting algorithm for a PRAM used a probabilistic technique.

Proposition 3.69 [Re81]
A CREW PRAM with $p = n$ processors sorts n elements from a linearly ordered set S in average time $O(\log n)$, using a probabilistic algorithm.

3.6 Distributed systems models

The multiprocessor machine model discussed in Section 3.4 is characterized by a global memory. Single processors can use this memory, according to a given protocol, to exchange locally computed results, to synchronize, and to store global data structures. This model is certainly a powerful theoretical tool. More debated however is the problem of whether it can be really transferred to a concrete architecture with massive parallelism. Indeed, machines with a large number of processors have a *distributed memory architecture*. This term refers to a parallel machine without a global memory, where each processor has exclusive access to a local memory. Access to data stored in the memory of another processor takes place by direct communication between the two processors involved. In this section we will briefly discuss an abstract model of a distributed system called *fixed connection network* (or simply *network*).

A network is an unlimited set of processors $P_0, P_1, ...$, each having a local memory consisting of an infinite sequence of registers: $r_0, r_1, ...$. In contrast to the PRAM, a global memory structure common to all processors is absent. Instead, direct connections are present allowing a processor to read data from (write data to) the local memory of a predefined set of processors.

Two processors linked by a direct connection are called *adjacent*. Each processor is uniquely identified by an integer number available in a local memory register called PIN.

All processors execute the same program and are globally synchronized. Regarding the instruction set the same considerations given in Section 3.4.2 on PRAMs apply. Adopting a unit cost criterion inhibits the use of too powerful instructions. The minimal instruction set guarantees that word size $W(n)$, that is, the maximum number of bits needed to represent data, memory addresses or processors during the computation, does not grow "too much" with respect to the initial values (input data, value of n and the maximum number $P(n)$ of processors used in the computation). If the above is guaranteed by the computation itself, it is then possible to use richer instruction sets without violating the theses of sequential and parallel computation.

Besides the instructions of a classical PRAM, the network model set must also include communication instructions, that is, instructions to read from/write to the memory of other processors. To this end, each processor has a second set of read-only registers, $p_0, p_1, ...$, acting as communication ports. Let us assume that gate p_k of processor P_i contains the value j. With this assumption, the execution of the instruction:

$x \leftarrow (y$ **of processor** $k)$

by P_i determines the transfer (read) of the contents of register r_y of processor P_j to register r_x of P_i. With the same assumption, the execution of the instruction:

$(y$ **of processor** $k) \leftarrow x$

by P_i determines the transfer (write) of the contents of register r_x of processor P_i to register r_y of P_j.

It is obvious then that the criteria adopted in the PRAM case can also be adopted in the case of networks. We will then have EREW, CREW and CRCW networks (with further differences in the case of a CRCW network).

The fact that the communication gate k of processor P_i contains the value j is clearly an abstraction. In reality this means that processors P_i and P_j are linked by a *unidirectional* line, and that the communication ports, which constitute the terminals of such a line, are interface registers (see Figure 3.24).

A network computation is defined in a similar way to a PRAM computation. The main difference is in the input/output protocol. In a computation using $P(n)$ processors, the ith input datum will be stored in

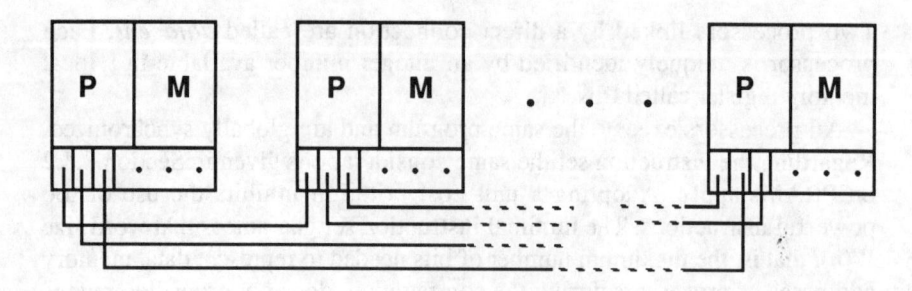

Figure 3.24 Processor network.

register $r_{\lfloor i/P(n) \rfloor}$ of processor $P_{i \bmod P(n)}$. A similar convention can be adopted for the output of length m.

All processors are active at the beginning of the computation. Furthermore, for any $i \in [0, P(n) - 1]$, the processors with which P_i can communicate (that is, the values contained in the communication ports of P_i) are determined before the beginning of the computation. In general, the maximum number of processors adjacent to any processor of the network depends on the value of n. Alternatively, one can adopt an activation scheme similar to the one originally defined for the PRAM (with only one initially active processor), and expect that the adjacent processors are also dynamically determined (see [GP81]).

Until now, the notion of a distributed system has been presented in an informal way. From a formal point of view, a network is an interconnection structure and a finite-length program (see [GP81, Pa86]).

Definition 3.70
The *interconnection structure* of a network is a triplet $I = (P, D, G)$, where:

- $P:N \to N$, and $P(n)$ is the *number of processors* which can be used in a computation on input of size n.
- $D:N \to N$, and $D(n)$ is the *degree of interconnection* of the network, relative to a computation on input of size n. The degree of interconnection is the maximum number of processors adjacent to any processor of the network.
- $G:\{i: 0 \leq i < P(n)\} \times \{d: 0 \leq d < D(n)\} \times N \to \{i: 0 \leq i < P(n)\}$ is the *interconnection function*. For any $0 \leq d < D(n)$, $G(i, d, n)$ is the processor connected, by means of the gate p_d, to processor P_i, $0 \leq i < P(n)$. If $G(i, d, n) = j$, P_j is said to be *adjacent* to P_i.

Definition 3.71
A network with fixed interconnection structure is a pair $R = (I, M)$, where I is an interconnection structure and M is a finite-length program.

Given n, the function G defines a direct graph labelled H, called *interconnection graph*. H has $P(n)$ nodes, representing the processors, and arcs (i, j) such that $G(i, d, n) = j$ for a $d \in [0, D(n) - 1]$. In contrast to the PRAM, where each processor can communicate with all the others through the global memory, in a network a processor can communicate *directly* with at most $D(n)$ other processors, namely the adjacent ones in the interconnection graph.

In the case of families of circuits, we required a uniformity condition allowing the description, using a finite amount of information, of every element of the family, one for each value of the size n. In the PRAM case the required condition was that the program, of finite size, be unique for each value of n. With networks we have a problem similar to that of uniform circuit families. We want the interconnection structure to be computable, and with a reasonable computational effort. In fact, as in the circuit case, if we simply assume that the function G (determining, for any n, the network topology for computations on inputs of size n, for any $n \in N$) is computable, we noticeably weaken the model as a tool to design algorithms. Consider, in this respect, the following example.

Example 3.72 [Pa87]
Let us consider any Boolean computable function f, that is, a function $f:\{0,1\}^n \to \{0,1\}$. A network with $P(n) = n2^n$ processors, with a complete instruction set and a weak CRCW conflict resolution scheme, can compute f in constant time. To this end, let us divide the $P(n)$ processors in 2^n groups of n processors each. The first processor in each group, with index in, $i = 0$, 1, ..., $2^n - 1$, will be called group *manager*. As is well known, there are 2^n possible configurations of the input values. The task of the processors from the ith group is to recognize a given configuration of the input bits, namely the configuration which, interpreted as a base 2 number, corresponds to the value i. This can be obtained if the jth processor of the ith group compares the jth input bit with the jth bit of the binary representation of the number i ($0 \le j < n$). When the two bits are different, the processor writes the value 0 to a suitable variable, say f, of its manager. Let us assume that the content of f has been initialized with the value 1 in every group. In this case, if the input configuration value is k, at the end of the comparison phase only the manager of the kth group will find the variable f still with value 1. Up to this point the interconnection structure is very simple. Each processor is connected to the manager of its own group. Furthermore, the jth processor in each group is connected to processor P_j, containing the jth input bit, $0 \le j < n$. Assume now that the function f has been previously computed for every possible input configuration. Let us then connect the manager of the ith group with

processor $f(i)$. Formally, this means that a communication gate, let it be the sth ($0 \leq s < D(n)$), of processor P_{ni} is initialized with the value $G(ni, d, n) = f(i)$. In this way, the manager of the kth group can give the value $f(k)$ by simply reading the PIN from the processor it is connected to by means of gate s. The program executed by each processor is given below. The jth input bit is stored in the variable x of processor P_j, $j = 0, ..., n$.

begin
    $i \leftarrow \lfloor \text{PIN}/n \rfloor$;   **comment** computing the group number;
    $j \leftarrow \text{PIN} - i * n$;   **comment** computing the number within the group;
    if $j = 0$ **then** $f \leftarrow 1$;
    $temp \leftarrow i * 2^{-j}$;
    $jb \leftarrow temp - \lfloor temp/2 \rfloor * 2$;   **comment** jb is the j-th bit of i;
    if (x **of processor** j) $= jb$ **then begin**
        if $j = 0$ **then** $f \leftarrow 0$
        else (f **of processor** $i * n$) $\leftarrow 0$ **end**;
    if ($j = 0$) **and** ($f = 1$) **then**
        (*output* **of processor** 0) $\leftarrow$ (PIN **of processor** $f(i)$)
end

The previous example represents a limiting case. The interconnection structure is indeed simply used as a table. The values in the table are the possible values taken by the function. The "computation" is then reduced to accessing the table position where the value corresponding to the input configuration is found.

This is certainly not the intended use of a network. In the example one is somewhat "cheating". The low computational cost is obtained at the expense of a high cost in defining the interconnection structure, which requires in fact computing the function f for any possible input configuration. Note then that, as in the case of families of circuits, the cost of defining the interconnection structure gives in some sense the cost of producing the structure itself. If the amount of computational resources is exceedingly high, it is impossible to automate the production of circuits or of networks. The following uniformity requirement, suggested in [GP81], prevents anomalous situations as given in the example.

Definition 3.73
A network is said to be uniform if, for any n, and for any computation with $P(n)$ processors, the function $G(i, d, n)$ can be computed by a deterministic Turing machine using $O(\log P(n))$ space and $O((\log P(n))^2)$ time.

The *distance* between two processors is defined as the distance of the corresponding nodes in H. If $D(n) < P(n)$, it is possible that during the computation the need arises for one or more processors to communicate with processors not directly connected with them. These communications must be divided into one or more communications between adjacent processors, some of which act simply as intermediaries. Being implemented by a single instruction, communication between adjacent processors has unit cost. Conversely, communication between non-adjacent processors will have a cost at least equal to the distance between sender and receiver.

The above arguments point in the direction of adopting large values of the interconnection degree, ideally $D(n) = P(n)$. On the other hand, an extra arc in the interconnection graph means an extra physical connection in the network. This will cause an increase in system cost. This cost includes, besides the cost of the connections proper, the cost stemming from a greater complexity in the logic used to control connections and transmissions. Note, in this respect, that the number of connections needed to keep the interconnection graph complete increases with the square of the number of processors. Hence a network where $D(n) = O(P(n))$ is not cost effective. It is even unreasonable as a theoretical model for the design of highly parallel algorithms, or for the study of lower complexity bounds. Indeed, the PRAM and the circuit models, respectively, are certainly better suited to these purposes. Conversely, if the degree of interconnection is limited, the network model is an interesting alternative, since in this case it describes in a formal way the characteristics of machines which can be implemented in practice. In this perspective, the study of efficient algorithms for the network model has a relevant practical importance.

We will say that the degree of interconnection of a network is bounded when $D(n) = O(1)$, or if it is a slowly increasing function of n, $D(n) = O(\log n)$ for instance (see [VB81]). Unluckily, a network with a limited degree of interconnection cannot efficiently compute any function. It has been proven in [GP81] that there are two structures, G_0 and G_1, and two functions, f_0 and f_1, such that a network with structure G_i ($i = 0, 1$) can efficiently compute (in $O(\log n)$ time) the function f_i, but it cannot efficiently compute (that is, it requires a time $\Omega(n)$) the function f_{i-1}.

There is a vast literature concerned with finding both interconnecting structures (and therefore graphs) providing efficient algorithms for well-defined classes of problems, and structures allowing an efficient emulation of the greatest possible number of other structures.

Some specific interconnection networks will be described in Section 5.2.

Bibliographic notes

The literature on models and, more generally, on the complexity of parallel computation is already extensive, although this field is still very active.

The book by Savage [Sa76] is an excellent introduction to the theory of Boolean circuits, which were already studied before the age of parallel computers. The works by Borodin [Bo77], Pippenger and Fischer [PF79] and by Pippenger [Pi79] establish the fundamental relations between sequential and parallel complexity measures. The most important work on the problem of circuit uniformity is that by Ruzzo [Ru81]. Finally, let us recall the work by Cook [Co85] which presents, besides the main properties of the circuit model, a wide list of "fast" circuits to solve important computational problems. Specialized references are also found in these papers.

The literature on PRAMs is even more extensive. The original paper by Fortune and Wyllie [FW78] has been discussed in depth in this book. The best way to start gaining a deeper insight is to read the reviews by Karp and Ramachandran [KR90] and by Eppstein and Galil [EG88], going on later to the references therein.

The basic reference for a further study of the theory of arithmetic circuits is the paper by von zur Gathen [Ga86]. The main model for comparison problems is the PRAM. However, the parallel comparison tree has been used in several papers, among which are [Va75] and [BH82]. Also, recall the work by Borodin [Bo82] which presents interesting observations on the relations between general and structured models (among which are arithmetic circuits and models for comparison problems).

The literature on distributed systems models is also substantial. We recommend, in particular, the recent book by Leighton [Le91]. For the more specific modeling aspects one can read the papers by Valiant and Brebner [VB81] and by Galil and Paul [GP81]. Also, the book by Parberry [Pa87] is at the introductory level.

A review paper on complexity, considering all the main models, is that by Cook [Co81].

Finally, we remind readers that most of the relevant results have been presented at the ACM Symposium on Theory of Computing conferences and at the IEEE Symposium on Foundations of Computer Science.

Problems

3.1 Design an $O(\log n)$ depth, $O(n)$ size Boolean circuit family that computes the following function: $f = \langle f_n \rangle_{n \in N}$, where:

$f_n: \{0,1\}^n \to [0, \lceil \log(n+1) \rceil]$

and $f_n(k)$ is the number of 1s in the n-bit input k.

3.2 A formula is a Boolean circuit with fan-out 1 (i.e. a circuit in which all the operation nodes have fan-out 1). Let FORMULA-SIZE($T(n)$) denote the set of languages that can be recognized by formulas of size

$T(n)$. Show that $\mathrm{NC}^1 \subseteq \mathrm{FORMULA\text{-}SIZE}(n^{O(1)})$. (Hint. Show that replicating k times the subcircuit from the inputs down to a node with i with fan-out $k > 1$ does not substantially increase the size.)

3.3 This is the inverse of Exercise 3.2. Prove that $\mathrm{FORMULA\text{-}SIZE}(n^{O(1)}) \subseteq \mathrm{NC}^1$. Note that, together with Exercise 3.2, this implies $\mathrm{NC}^1 = \mathrm{FORMULA\text{-}SIZE}(n^{O(1)})$. (Hint. Observe that a "balanced" formula with $n^{O(1)}$ nodes has $O(\log n)$ depth. Therefore you must show that anything that can be computed by a formula with $n^{O(1)}$ nodes can also be computed by a balanced formula with $n^{O(1)}$ nodes. Try to write a balancing procedure.)

3.4. Design an $O(\log n)$ depth circuit family for integer addition based on the following idea. For $i = 2, \ldots, n$, there is a carryover from the $(i-1)$th position if:

(a) carryover has been generated exactly on the $(i-1)$th position, or
(b) carryover has been generated on the jth position (where $j < i - 1$) and propagated through the positions $j + 1, \ldots, i - 1$.

3.5 Show that two $n \times n$ matrices can be multiplied in a time $O(\log n)$ on an EREW PRAM.

3.6 Discuss a few theoretical consequences on complexity of the following (hypothetical) improvement of Corollary 3.52:

$$\mathrm{TP}_{\mathrm{EREW}}(\log^k n, n^{O(1)}) \subseteq \mathrm{TP}_{\mathrm{str}}(\log^k n, n^{O(1)}), \; k \geq 1$$

3.7 Prove that any Boolean function $f:\{0,1\}^n \to \{0,1\}$ can be computed by an arithmetic circuit.

3.8 Design an EREW PRAM algorithm that, given a linked list L with $2n$ elements, returns two lists, of n elements each, consisting of the elements in even and odd position in L respectively. Give a scheme for allocating the list elements so that your algorithm can run in less than linear time. (Hint. Assign different elements to different processors, assuming that the number of processors is at least $2n$.)

Chapter 4
Parallel Algorithms

This chapter is devoted to algorithms for parallel models of computation. Note that, as opposed to the sequential case, the evaluation of parallel algorithms strongly depends on the model used. Because of the differences among models, it is useful to specify the model being referred to in order to avoid ambiguities.

In this chapter we will first introduce some general techniques (Section 4.1). We then will consider some arithmetic algorithms (Section 4.2), combinatorial algorithms (Section 4.3) and numerical algorithms (Section 4.4). We will evaluate algorithms in different models: arithmetic algorithms will be analyzed with respect to Boolean circuits, combinatorial algorithms with respect to PRAM, and numerical algorithms in terms of time and processor number of the algorithmic model and/or of arithmetic circuits.

4.1 Basic techniques

The most natural approach to developing parallel algorithms is to start from a sequential algorithm, then to identify the independent steps, and finally to obtain a parallel version (see Section 3.2). As we already mentioned in the Introduction, a simple conversion from sequential to parallel often produces algorithms with a low degree of parallelism. Hence there arises the problem of determining strategies for obtaining efficient parallel algorithms. There is no general method for the best parallelization of a given problem. However, there are techniques that are successfully used in different problems.

In this section we will introduce a few widely applicable methods. In this section, we denote the resources time, hardware, number of processors and simultaneous time–number of processors with T, H, P and PT respectively.

4.1.1 Divide et impera
Divide et impera is a well–known technique in sequential computing. It consists of dividing the problem into subproblems of the same kind but of

lesser size and then rearranging the solutions. With parallel computations, this technique is sometimes very efficient. In fact, it allows us to express problems in terms of independent subproblems. These in turn can be solved in parallel, and their solutions must then be combined to obtain the solution to the original problem. The last additional part can be done only after all subproblems have been solved. Let us formalize the cost of the parallel solution of a computational problem of order n, in terms of size $H(n)$ and depth $T(n)$ (see Chapter 3). Let r be an integer dividing n, and let $H_i(n/r)$ and $T_i(n/r)$, $i = 1, 2, ..., r$, be the costs of solving r subproblems. Moreover let $T_c(n)$ and $H_c(n)$ be the additional (parallel) cost of solving the original problem after its subproblems have been solved. We have:

$$T(n) \leq \max\{T_1(n/r), T_2(n/r), ..., T_r(n/r)\} + T_c(n) \qquad (1)$$
$$H(n) \leq \max\{H_1(n/r) + H_2(n/r) + ... + H_r(n/r), H_c(n)\} \qquad (2)$$

Relation (1) follows from two facts:

- the independence of problems;
- the need to execute the additional work only when the subproblems have been solved.

Relation (2) follows from two other facts:

- each subproblem, being solved in parallel with the others, needs its own hardware resources;
- the total amount of hardware is the maximum of the resources needed to solve all subproblems, and the resources needed to obtain the final solution to the original problem.

Under the hypothesis that all subproblems are equivalent in terms of parallel cost (this assumption is often true in practice), (1) and (2) become:

$$T(n) \leq T(n/r) + T_c(n) \qquad (3)$$
$$H(n) \leq \max\{r\, H(n/r), H_c(n)\} \qquad (4)$$

where $T(n/r)$ and $H(n/r)$ are the cost of the subproblems of order n/r.

Relations (1) – (4) are very useful for analyzing parallel algorithms according to the *divide et impera* method.

An important example of (3) is when $T_c(n) = O(\log^k n)$, $k = 0, 1, ...$ In such a case the solution is $T(n) = O(\log^{k+1} n)$. This means that when $T_c(n) = O(\log^k n)$, $k = O(1)$, that is, *divide et impera* produces circuits with polylogarithmic depth.

4.1.2 Vectorization

Vectorization is a technique that, starting from the analysis of a problem, produces algorithms working on data structures well suited for vector computations (typically vectors and matrices). This technique is particularly important in SIMD (see Section 1.3), especially when it is convenient to work on a vector of order n instead of n scalars. Some examples of vectorization will be shown in Section 5.1.

4.1.3 Vector iterations

The method of vector iterations has been introduced by Traub and others for solving tridiagonal linear systems [Tr73, HST76]. Starting from a direct sequential method, a parallel iterative method is built.[1]

If we consider the recursion for computing the factorization LDU of a tridiagonal matrix, i.e:

$$x_1 = a_1$$
$$x_i = a_i - \frac{b_i c_i}{x_{i-1}}, \ 2 \leq i \leq n$$

it is possible to obtain the following iterations:

$$x_i^{(0)} = a_1$$
$$x_i^{(j)} = a_i - \frac{b_i c_i}{x_{i-1}^{(j-1)}}$$

where j is the iteration number. These iterations can be vectorized, i.e. they can be considered as updates of vector $\mathbf{x}^{(j)}$ with elements $x_i^{(j)}$, $j \geq 0$. In fact, at the first step ($j = 0$) each element of vector $\mathbf{x}^{(0)}$ contains the value a_1. At the second step the ith element of $\mathbf{x}^{(1)}$ contains the value $a_i - \frac{b_i c_i}{x_{i-1}^{(0)}} = a_i - \frac{b_i c_i}{a_1}$, $i = 1, 2, ..., n$. All the elements of $\mathbf{x}^{(1)}$ can then be updated simultaneously. The same principle applies to the subsequent iterations.

4.1.4 Recursive doubling

Recursive doubling consists of turning a computation graph such as the one shown in Figure 4.1(a) into the graph shown in Figure 4.1(b). It enables us to obtain a graph with logarithmic depth from a graph with linear depth. Let us consider the following example.

---

[1] Methods for solving linear systems can be divided into two classes: direct and iterative methods. Direct methods would compute the exact result if there were no data representation or arithmetic errors. On the other hand, iterative methods would anyway introduce an approximation error.

Basic techniques   127

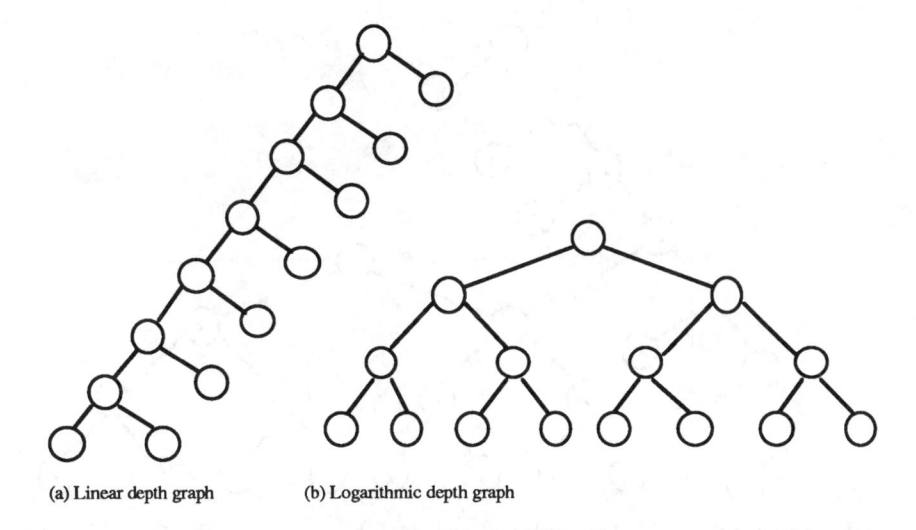

(a) Linear depth graph   (b) Logarithmic depth graph

Figure 4.1 $n = 8$. (a) computation graph with depth 7;
(b) computation graph with depth 3.

Example 4.1
Let us consider the computation of x^n, where $n = 2^k$, for a positive integer k.
Consider the algorithm:

begin
    $s \leftarrow x$;
    for $i \leftarrow 1$ **until** $n - 1$ **do**
    $s \leftarrow s*x$;
end

A computation graph with linear depth (see Figure 4.2(a)) corresponds to this algorithm. If the computation is organized in the following way:

begin
    $s \leftarrow x$;
    for $i \leftarrow 1$ **until** k **do**
    $s \leftarrow s*s$;
end

we obtain the graph in Figure 4.2(b), which has a logarithmic depth.

Usually, computation graphs of sequential algorithms for arithmetic computations have the structure shown in Figure 4.1(a). It is not always obvious whether the dependence among the operations is due to the algorithm or intrinsic to the problem. In the latter case, one cannot do much. In the first

(a) Linear depth graph **(b)** Logarithmic depth graph

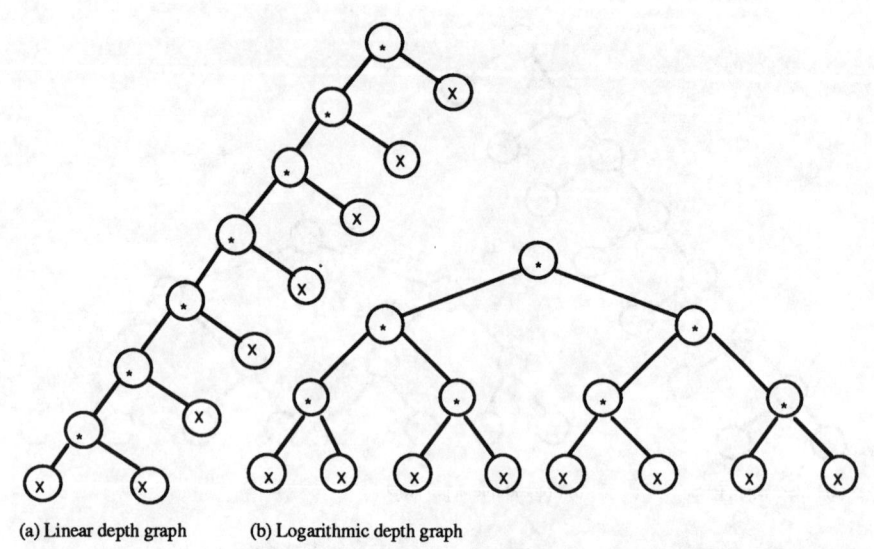

Figure 4.2 $n = 8$. (a) computation graph for the sequential computation of x^8;
(b) computation graph for the parallel computation of x^8.

case, an accurate analysis of the problem can reveal a possible recursive doubling approach.

Recursive doubling could also be considered as a special case of *divide et impera* (Section 4.1.1). In fact, the graph in Figure 4.2(b) suggests that the value x^n equals the product $x^{n/2}*x^{n/2}$; this in turn means that the problem P_n, i.e. the computation instead of x^n, can be decomposed as $P_{n/2}*P_{n/2}$, rather than $P_{n-1}*x$.

More generally, recursive doubling consists of obtaining a parallel algorithm with recursions such as (3) and (4), starting from a sequential algorithm, with cost $S(n)$ given by:

$$S(n) \leq S(n-r) + f(n)$$

where r is a constant and $f(n)$ is a non–decreasing function of n. The rightmost term of these recursions corresponds to the cost of solving subproblems of dimension n/k, $k = O(1)$. The term doubling is used because it is often $k = 2$, so that starting from the solution of a given problem, it is possible to obtain the solution of a problem of twice the size.

4.1.5 Reducing the number of processors

Sometimes it is possible to reduce the order of magnitude of the number of processors used by a parallel algorithm by increasing the execution time by a constant factor. Let us see how this goal can be attained.

Assume that a parallel algorithm takes time O(log n) and uses O(n) processors. Assume further that the best available sequential algorithm (solving the same problem) has a linear cost. It follows that the parallel algorithm has efficiency O(1/log n), which is far from the optimum. Sometimes it is possible to transform a parallel algorithm with such performance into another parallel algorithm with time O(log n) and number of processors O(n/log n), hence with efficiency θ(1). We proceed as follows. The problem of size n must be divided, according to divide et impera, into n/log n subproblems of size log n. Each subproblem is then solved by a single processor using the best available sequential algorithm (linear with log n). If the solution of the n/log n subproblems gives rise to the initial problem of order n/log n, we can use the parallel algorithm, taking time O(log(n/log n)) = O(log n) with n/log n processors.

In general, the following equality must apply to a problem P_n:

$$P_n = k\, P_{n/k} + P_k \tag{5}$$

for k dividing n.

Relation (5) means that problem P_n can be partitioned into k subproblems $P_{n/k}$ of the same type. Their solution originates problem P_k, which can be solved after all k subproblems have been solved.

Assume that in order to solve P_n, there exist:

- a parallel algorithm A_p with performance $T_{A_p}(n)$ and $H_{A_p}(n)$;
- a sequential algorithm A with cost (i.e. number of operations) $T_A(n)$.

Assume further that $T_{A_p}(n) H_{A_p}(n) = T_A(n) f(n)$, where $f(n)$ is a function increasing with n. The efficiency of A_p is thus given by $1/f(n)$, and this means that the algorithm makes a redundant use of resources.

We can then consider a function $g(n)$ such that:

1. $\Omega(1) = g(n) < n$;
2. $T_{A_p}(g(n)) H_{A_p}(g(n)) = T_A(n)$.

Replacing k with $g(n)$ in (5) we obtain:

$$P_n = g(n)\, P_{n/g(n)} + P_{g(n)} \tag{6}$$

According to (6) it is possible to solve P_n as follows:

1. solve $P_{n/g(n)}$ with a single processor using algorithm A;
2. solve $P_{g(n)}$ using the parallel algorithm A_p.

This procedure corresponds to a new parallel algorithm A^* with performance T^* and H^* given by:

$$T^*(n) = T_{A_p}(g(n)) + T_A(n/g(n))$$

and:

$$H^*(n) = \max\{H_{A_p}(g(n)), g(n)\}$$

The efficiency of A^* is then $O(1)$, provided that the following equalities apply to $g(n)$:

$$g(n) = O(H_{A_p}(g(n)) \text{ and } T_{A_p}(g(n)) = O(T_A(n/g(n)))$$

Let us see an example where this technique is applied.

Example 4.2 (computing the sum $s = a_1 + a_2 + ... + a_n$, with $n = 2^k$)
The fan–in algorithm for computing s takes time $O(\log n)$ and hardware $O(n)$.
Let $s_i = a_{(i-1)k+1} + a_{(i-1)k+2} + ... + a_{(i-1)k+k}$, $i = 1, 2, ..., 2^k/k$. Thus $s = s_1 + s_2 + ... + s_{n/\log n}$. It is then possible to apply the technique discussed above:

1. compute s_i within processor i in time k, $i = 1, 2, ..., 2^k/k$;
2. compute $s = s_1 + s_2 + ... + s_{n/\log n}$ using the fan–in algorithm.

It is easy to verify that we obtained a parallel algorithm with constant efficiency (i.e. an optimum algorithm), starting from an inefficient algorithm. A similar technique will also be used in Section 6.4.

We have seen how under certain conditions it is possible to reduce the number of processors. The first condition is the possibility of describing the problem according to (5). It is possible to generalize (5), i.e. to obtain a problem P_n for which the equality:

$$P_n = k P_{n/k} + Q_k$$

applies for some k dividing n, where Q is not necessarily the same problem as P. Also, in this case we can perform an analysis similar to the one derived from (5).

4.2 Arithmetic algorithms

This section discusses parallel algorithms for solving some of the main arithmetic problems, e.g. computing the sum, the product and the quotient of two integer numbers in binary representation. Section 4.2.1 presents a circuit family in NC^1 for integer addition. Section 4.2.2 describes two methods for computing the product of integers, in logarithmic time, while Section 4.2.3 analyzes the problem of integer division.

4.2.1 Adding integer numbers

The task of the parallel computation of the sum of two numbers presents only the problem of computing the carry that can propagate through all digits. This is why the crucial point is to determine an efficient method for computing the carry in parallel. In the following we present a solution to this problem.

Let $X = x_{n-1}x_{n-2} \ldots x_0$ and $Y = y_{n-1}y_{n-2} \ldots y_0$ be two n-bit numbers to be added. Let $g_k = x_k \wedge y_k$ and $p_k = x_k \vee y_k$, $k = 0, \ldots, n-1$. g_k is called the kth *carry-generate bit*, whereas p_k is called the kth *carry-propagate bit*. It is obvious that both carry-generate and carry-propagate bits can be computed with constant depth and linear size. Using these bits we can write the following recursion to compute the kth carry bit: $c_k = g_k \vee (p_k \wedge c_{k-1})$. By iterating the recursion, we obtain $c_k = g_k \vee (p_k \wedge g_{k-1}) \vee (p_k \wedge p_{k-1} \wedge g_{k-2}) \vee \ldots \vee (p_k \wedge p_{k-1} \wedge \ldots \wedge p_1 \wedge g_0)$. This logical expression can be computed by a circuit with depth not greater than $2\lceil \log k \rceil$ and size k^2.

Once the values c_k, $k = 0, \ldots, n-1$ have been obtained, it is possible to compute the result $Z = z_{n-1}z_{n-2} \ldots z_0$ with constant depth and linear size through the expression $z_k = x_k \oplus y_k \oplus c_{k-1}$, $k = 0, \ldots, n-1$. The whole circuit has depth $O(\log n)$ and size $O(n^3)$, hence it is in NC^1 (the uniformity can be easily proven).

The addition problem will be considered again in Section 6.4.1, where we will illustrate a particularly efficient VLSI circuit corresponding to a Boolean circuit with depth $O(\log n)$ and size $O(n)$ (see Section 3.3).

4.2.2 Multiplying integer numbers

Many researches have been dedicated to the development of efficient and fast methods for multiplying two integer numbers in parallel. Owing to the great importance of this problem, both theoretical and practical algorithms for fast hardware implementations have been investigated. The multiplication of integers is in NC^1, according to many proposed algorithms. We will hereafter illustrate two algorithms dating back to the sixties, which produce logarithmic depth circuits.

The first scheme is the *Wallace tree* [Wa64]. It originates from the analysis of the methods often used to multiply two numbers, resembling what we learn in primary school, where the multiplication of an n digit number by an m digit number is performed by m additions of $m + n$ digit numbers (see Figure 4.3).

Wallace observed that this scheme does not consider the particular structure of the operands, each one being a simple multiple of the multiplicand. Wallace multipliers are based on the acceleration of addition obtained by using *pseudoadders*.

132 Parallel algorithms

$$7845573 \times 1567 = \qquad 7845573 \times 7 \;+$$
$$78455730 \times 6 \;+$$
$$784557300 \times 5 \;+$$
$$7845573000 \times 1$$

Figure 4.3 Classical multiplication algorithm (base 10 numbers).

Pseudoadder devices were used in the sixties. They receive three numbers as input and produce a pair of numbers with the same sum as the three input numbers. Pseudoadders are much faster than adders because they do not have the problem of carry propagation. A pseudoadder can be made in several ways. As an example, a circuit adding two numbers (see Figure 3.9) can be used: just replace the inputs normally used for receiving carries with the third input number, and the outputs normally used for producing carries with the second output number.

The Wallace algorithm can be represented as a tree structure where each node is a pseudoadder with three input and two output numbers (see Figure 4.4). At each level the numbers to be added are divided into groups of three. Each group is the input to a pseudoadder, hence the number of operands increases by a factor of 1.5 when passing from one level to the next in the tree. The number of levels is then $O(\log n)$ and the total depth is $O(\log n)$. The hardware used is polynomial in n.

To be more precise, if D and W are the depth and the width of the Wallace tree, then:

$$D(n) \leq D(2n/3) + O(1)$$
$$D(2) = D_A(n)$$

where D_A represents the depth of an adder circuit, whereas:

$$W(n) = 2n/3 \; W_p(n)$$
$$W_p(n) = W_A(n)$$

where W_A and W_p represent the width of an addition and a pseudoadder circuit, respectively.

It follows that:

$$D(n) = O(\log n + D_A(n)) \text{ and } S(n) = O(n \; W_A(n))$$

Hence if an adder is used such that $D_A(n) = O(\log n)$ and $W_A(n) = O(n)$, the performances of the Wallace multiplier are given by:

$$D(n) = O(\log n) \text{ and } W(n) = O(n^2)$$

Arithmetic algorithms 133

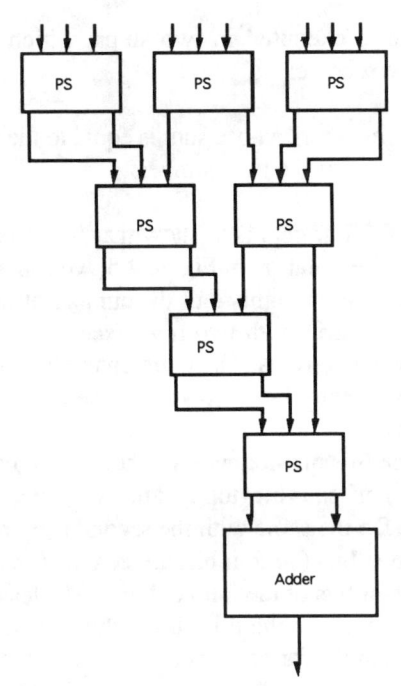

Figure 4.4 Wallace tree for adding nine numbers, where PS is used to denote a pseudoadder.

Let us now consider another algorithm introduced by Dadda in 1965 [Da65].

Assume that the numbers are in binary format. The algorithm uses *parallel counters* (n, m), i.e. devices that, given an n-bit input number, produce the number of bits equal to 1 in the input. The output is then at most n, hence it can be represented with $m \leq \lceil \log n \rceil + 1$ bits. The multiplication problem is described by means of a matrix (see Figure 4.5) representing the classical multiplication algorithm.

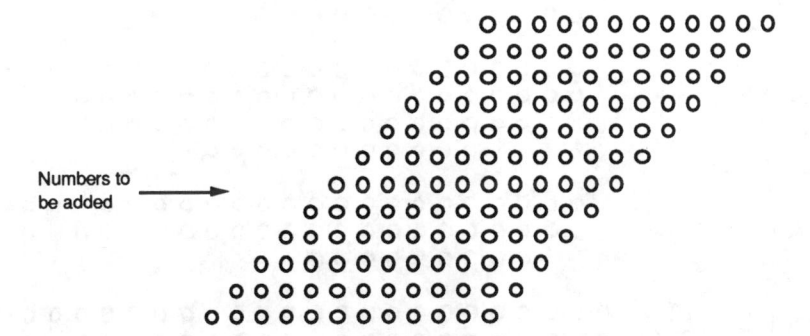

Figure 4.5 Representing the problem by means of a matrix (case $n = 12$).

134   Parallel algorithms

The multiplication is executed in two steps, which is similar to what happens in the Wallace scheme:

1. compute two numbers whose sum is equal to the desired product;
2. compute the sum of the two numbers.

Since addition is in NC^1, our goal is to show that step 1 is in NC^1.

Starting from the representation in Figure 4.5, we can generate a sequence of matrices, each one having roughly half the number of rows of the previous one, until we obtain a matrix with two rows (see Figure 4.6). The product will be the sum of these two rows. The main characteristic of this method is to avoid carry propagation until the last stage, where two numbers are added using a complete adder.

Let us now describe this method with the help of Figure 4.6. Simply start by adding the least significant bit (top rightmost corner in the figure) from matrix A to matrix B. Do the same with the second rightmost column, which is only composed of two bits (bottom bits are zeros). Apply then an adder (3, 2) to the three significant bits of the third column. The least significant output bit from the adder is written in the third last column of B, whereas the most significant bit is stored in the fourth last column and in the second row. Since

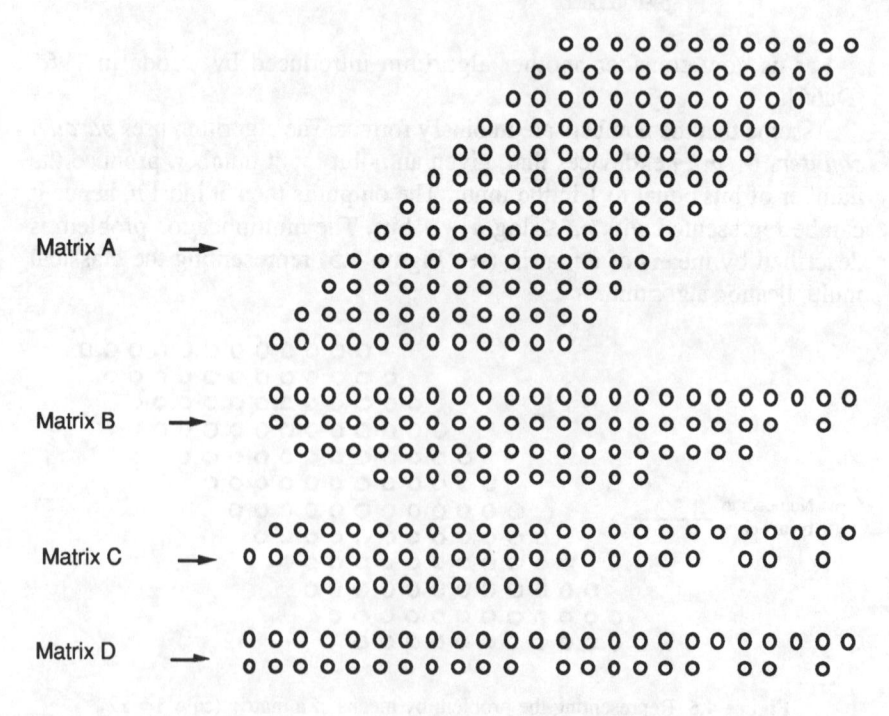

Figure 4.6 Representing the algorithm as a series of matrices (A, B, C, D) (case $n = 12$).

the fourth rightmost column of A has four bits, the corresponding counter must have three output bits: the least significant one is stored in the first row and in the fourth rightmost column of B. The following bit is then stored in the second row and in the fifth column of B, and the most significant bit is stored in the third row and in the sixth column. The procedure continues in a similar way, until all elements of B have been determined. The other matrices are then built in the same way.

The correctness of the algorithm can be proven according to the following observations: the product p to be computed is the weighted sum of all bits a_{ij} corresponding to elements of A, that is:

$$p = \sum_{j=1}^{2n-1} \sum_{i=1}^{n} a_{ij} 2^{j-1}$$

and the contribution given to p by column j of A is:

$$\sum_{i=1}^{n} a_{ij} 2^{j-1} = 2^{j-1} \sum_{i=1}^{n} a_{ij}$$

The same contribution to p is given by the jth column of B (see Figure 4.6), since $\sum_{i=1}^{n} a_{ij}$ is the output of the counter with the least significant bit in column j, i.e. its weight is 2^{j-1}. It follows that the matrices A and B are equivalent for computing p. The same property can be proven for the generated matrices. It is then easy to prove that the whole circuit has depth $O(\log n)$.

Many efficient algorithms for integer multiplication use the fast Fourier transform (FFT). In this way it is possible to reduce significantly the amount of hardware, while still maintaining a logarithmic time. The approach consists of interpreting integer multiplication as a discrete convolution, and of computing the convolution using an FFT algorithm.

We shall reconsider the problem of integer multiplication in Section 6.4.

4.2.3 Integer division

The problem of parallel division presents more difficulties than parallel addition and multiplication. Only recently an NC^1 circuit (P–uniform, but not log–space uniform) has been produced [BCH84]. Before this result, the fastest algorithm had been obtained by applying the Newton method [BBCM87] to the problem of computing the reciprocal of an integer, which is reducible to the integer division problem.

The Newton method consists of transforming, under appropriate conditions, an equation of the form $f(x) = 0$, where f is a function which can be differentiated at least twice, into the equation $x = x - f(x)/f'(x)$, hence the iterative method is obtained:

$x_0 = \alpha$
$x_{i+1} = x_i - f(x_i)/f'(x_i)$, $i = 0, 1, 2, ...,$

for a given α.

Given a real number $a \neq 0$, which we can suppose without loss of generality to be between 1/2 and 1, Newton iterations for computing the reciprocal of a can be obtained. Starting from equation $ax - 1 = 0$, whose solution is $1/a$ we obtain:

$x_i = x_{i-1} - x_{i-1}(ax_{i-1} - 1)$, $i = 1, 2, ...,$

where the value x_0 (called *starting point*) must be somehow determined. In order to have the convergence of the method, it is sufficient that $|ax_0 - 1| < 1$. Values of x_0 with this property can be easily determined.

Suppose that a is an n digit binary number, and that the problem is to compute n correct digits of $1/a$. It is then sufficient to perform $O(\log n)$ Newton iterations. This is both because the Newton method doubles the number of correct digits at each iteration (the method has a quadratic convergence) and because the starting point x_0 shares at least the first digit with $1/a$.

Since each iteration can be executed in time $O(\log n)$, as it consists of two multiplications and two additions of numbers of $O(n)$ bits, the Newton method produces an NC^2 circuit for computing the reciprocal of an n–bit number.

The following inequality applies:

Division $\leq_{NC^1}$ Reciprocal

from which we have that the Newton method produces an NC^2 circuit for division.

NC^1 P–uniform, but not log–space uniform, circuits for division have been suggested by Beame, Cook and Hoover [BCH84]. Their construction is very complicated and uses properties of prime numbers combined with the Chinese remainder theorem. The approach is based on reducing division to the computation of the reciprocal, and then reducing the latter to the computation of the iterated product. It produces an NC^1 (P–uniform) circuit for the iterated product, through which it is then possible to derive an NC^1 circuit for the division.

The following inequalities apply (see [BCH84] for the proof):

Division $\leq_{NC^1}$ Reciprocal $\leq_{NC^1}$ Powering $\leq_{NC^1}$ Iterated Product

The problem of computing the iterated product is expressed as follows:

given the n bits binary integers $a_1, a_2, ..., a_n$,
compute the n^2 bits integer number $a_1 a_2 ... a_n$

Let us see the main characteristics of the Beame, Cook and Hoover circuit for the computation of the iterated product $a_1 a_2 ... a_n$.

1. Let $p_1, p_2, ..., p_m$ be m distinct prime numbers smaller than n^2, the product of which is larger than 2^{n^2}. Further, let $p = p_1 p_2 ... p_m$.
2. Compute $a_i \bmod p_j$, $i = 1, 2, ..., n, j = 1, 2, ..., m$.
3. Compute $a_1 a_2 ... a_n \bmod c_j$, $j = 1, 2, ..., m$.
4. Compute $a_1 a_2 ... a_n \bmod c$, using the Chinese remainder theorem.

It is possible to prove [BCH84] that each one of the four previous steps can be implemented by circuits in NC^1, subject to the condition that suitable constants are precomputed and inserted in the circuit. This leaves doubts on the log–space uniformity of the circuit.

4.3 Combinatorial algorithms

This section considers some combinatorial algorithms, analyzing their performance with respect to the PRAM model (see Section 3.4).

4.3.1 Computing prefix sums

Let us first consider the problem of computing *prefix sums* [LF80].

Let @ be a binary associative operation, and let a_i, $i = 1, ..., n$, be elements to which @ can be applied. The prefix sums computation consists of computing the "sums" of prefixes $s_i = a_1 @ a_2 @ ... @ a_i$, $i = 1, ..., n$ (see Section 3.4.3).

The sequential solution can be obtained by simply applying $n-1$ times the operator @. A lower bound on the parallel time is $\lceil \log n \rceil$: this is due to the fact that operator @ has arity two, and that s_n depends on all n elements. We will then see that the algorithm that we are going to illustrate is optimum with respect to time. For the sake of simplicity, let us suppose that $n = 2^k$, with k positive integer.

Input: An array of n elements a_i stored in the global memory M_i, $i = 1, ..., n$. Assume that M_0 contains the value 0.
Working space: The local memory location L_n of processor P_i,
$i = 1, 3, ..., n - 1$.
if $n \neq 1$ **then**

Step 1 **for** i in $\{1, 3, ..., n-1\}$ **processor** P_i **executes** $L_n \leftarrow M_i$.
Step 2 **for** i in $\{1, 2, ..., n/2\}$ **processor** P_i **executes**
$$M_i \leftarrow M_{2i-1} @ M_{2i}.$$
Step 3 Recursively compute the prefix sums of $n/2$ elements in $M_1, ..., M_{n/2}$.
Step 4 **for** i in $\{2, 4, ..., n\}$ **processor** P_i **executes** $M_i \leftarrow M_{i/2}$.
Step 5 **for** i in $\{1, 3, ..., n-1\}$ **processor** P_i **executes**
$$M_i \leftarrow M_{(i-1)/2} @ L_n.$$

Note that step 2 is a recursive doubling (Section 4.1).

The operations specified in steps 4 and 5 (as well as those in steps 1 and 2) can be executed in parallel. Since all processors are synchronized, instructions such as the following:

$M_2 \leftarrow M_1$, executed by processor P_2;
$M_4 \leftarrow M_2$, executed by processor P_4;

do not interfere (i.e. the value used by P_4 is not that written by P_2, and so on).

It is easy to verify that the time $T(n)$ used by the described algorithm satisfies the relation:

$$T(n) = T(n/2) + O(1)$$

hence $T(n) = O(\log n)$.

As far as the number of processors $P(n)$ and the use of memory $M(n)$ are concerned, we have:

$$P(n) = O(n) \text{ and } M(n) = O(n)$$

It is easy to realize that the algorithm can be executed in time $O(\log n)$ on an EREW PRAM with $O(n)$ processors, as there are no conflicts on memory access. Moreover, the program specified by the algorithm can be uniformly generated in space $O(\log n)$. In fact, $O(n)$ processors and memory locations are used, so that $O(\log n)$ bits are sufficient to specify the steps of the program.

Finally, it is possible to modify the algorithm slightly in order to obtain an algorithm with time $O(\log n)$, using only $O(n/\log n)$ processors (see [KR90]).

4.3.2 List ranking

List ranking is a problem strictly related to the previous one:

given a list of n elements, compute the sums of the last i elements of the list, $i = 1, 2, ..., n$.

Hence it is a variant of the prefix sums problem, in which the elements belong to a list and the sums are computed starting from the end, not from the beginning. The term list ranking is usually used to represent a subproblem, i.e. when the value of each element is 1 and the operation @ coincides with addition. In this case, the solution provides the rank $r(i)$ of each element, i.e. the number of elements following it in the list.

Let us assume that the list is represented in terms of two n–vectors, $c = c(i)$ and $s = s(i)$, containing the elements and describing the function "successor" respectively. For each i, $c(i)$ contains the value of element i, whereas $s(i)$ gives the location of the successor of $c(i)$ in the list. Moreover, we can assume without loss of generality that there exists a special element z such that $c(i_n) = z$ and $s(i_n) = i_n$, where $c(i_n)$ is the last element of the list.

In the following we describe an algorithm solving the problem of list ranking in time $O(\log n)$ on an EREW PRAM of $O(n)$ processors.

begin
  for j←1 **until** ⌈log n⌉ **do begin**
    c(PIN):= c(PIN)@c(s(PIN));
    s(PIN):= s(s(PIN))
  end;
  write c(PIN);
end

The main operation of this algorithm is *pointer jumping*, i.e. replacing each pointer $s(i)$ with $s(s(i))$ (the pointer of the pointer). With this technique we obtain a double recursion, which allows us to solve the problem in time $O(\log n)$.

The correctness of the algorithm follows from the fact that at the beginning of each step, $c(i)$ is equal to the sum of the n elements of the list with rank $r(i), r(i) - 1, ..., r(s(i)) + 1$, depending on the current value $s(i)$. After ⌈log n⌉ iterations, for every i, we have $s(i) = i_n$.

If we assign each location i to a different processor, we obtain a time $O(\log n)$ on $O(n)$ processors, so that the efficiency of the algorithm is $O(1/\log n)$, since the best sequential algorithm uses a linear number of operations. Also, in this case some modifications can be introduced which allow us to obtain a constant efficiency.

4.3.3 Evaluating expressions

The problem of evaluating expressions can be formalized in the following way:

given an arithmetic expression **E**, with operations + and *, and with precedences represented by parentheses, evaluate **E** together with all its subexpressions.

Note that the computation of prefix sums is a particular case of an expression (with all the subexpressions) to be evaluated, i.e. $E = (...(a_1 + a_2) + a_3 ...) + a_n)$.

In this section we illustrate a method for evaluating expressions. As we will see, the associated computation graph is a binary tree.

It is possible to relate every expression with n pairs of parentheses to a binary tree with n leaves specifying the precedences. The value associated to each node is the value of the corresponding subexpression. In particular, the value of the whole expression is associated to the root node. If we compute the values of the intermediate nodes, starting from the leaves and going towards the root, a simple algorithm with linear cost is obtained. However, if the tree is not well balanced, this algorithm cannot be efficiently parallelized.

A method from which it is possible to derive an efficient parallel algorithm consists of *shrinking* the tree [MR85]: in $O(\log n)$ steps, a tree with n nodes is transformed into a tree consisting only of one node. It is possible to execute each step in constant time on a PRAM. The shrinking is applicable to *ordered* trees, where each vertex can be either a leaf or an intermediate node with two sons, whereas each arc connects a son f with its father $p(f)$. Hence the arc orientation is implicit.

The basic operation of the transformation process, called *shunt*, is described hereafter.

Shunt operation applied to a leaf f

Let f be a leaf of a binary tree A. A tree A' is built such that f and $p(f)$ are not present in A', while the other son f' of $p(f)$ has $p(p(f))$ as its father in A'.

An example of shunt operation is shown in Figure 4.7.
Let us describe the shrinking algorithm.

Input: an ordered binary tree A with n leaves.

(i) Label the leaves with 1, 2, ..., n, from left to right;
(ii) Repeat $\lceil \log n \rceil$ times the following steps:

(a) apply the shunt operation (in parallel) to all leaves which are left sons and which have odd numbers as labels;
(b) apply the shunt operation (in parallel) to all leaves which are right sons and which have even numbers as labels;
(c) remove the rightmost bit in the labels of all the other leaves.

Combinatorial algorithms   141

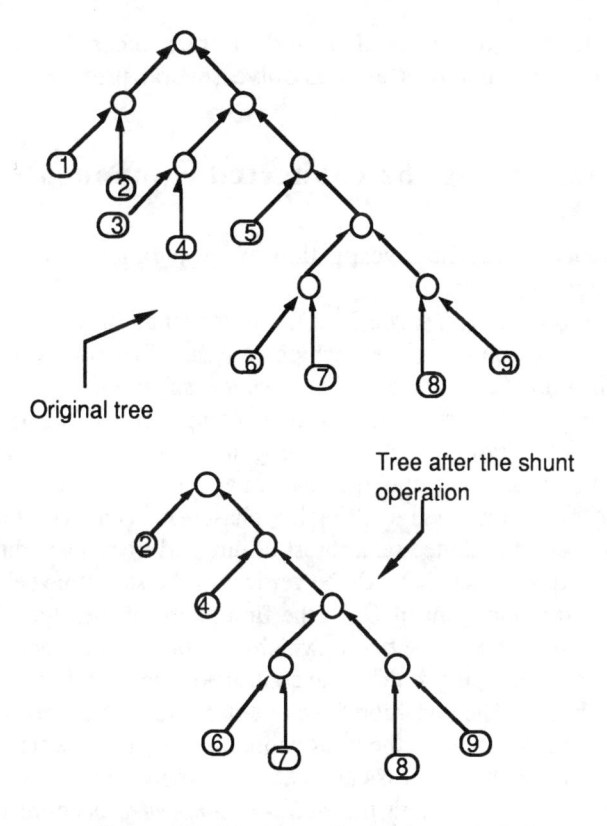

Figure 4.7 Shunt operation on leaves 1, 3, 5.

We still have to prove how it is possible to evaluate an expression through the shrinking algorithm. In order to do this, let us associate to each arc (u, v) of the tree an ordered pair of values (a, b): if the value of vertex u is x, then the arc (u, v) sends the operand $ax + b$ to vertex v. In order to ensure the correctness at the beginning of the computation we must associate the ordered pair $(1, 0)$ to each arc. Let us now consider a shunt operation executed on a leaf f with father p, brother s and grandfather q. Let v be the value associated to f and let (a_1, b_1), (a_2, b_2) and (a_3, b_3) be the values of arcs (f, p), (s, p) and (p, q). In the shrunk tree all these arcs are eliminated and substituted by arc (s, q). If a and b are two constants such that $ay + b = a_3((a_1v + b_1)*(a_2y + b_2)) + b_3$, where $*$ is the operation associated to vertex p, it is easy to verify that, assigning the value (a, b) to the arc (s, q), the shrunk tree represents the same expression as the original tree.

It is possible to prove that the described algorithm can be implemented in time $O(\log n)$ on an EREW PRAM.

The shrinking algorithm is also used in areas other than expression evaluation, in particular it is used to solve several problems on graphs [KR90].

4.3.4 Determining the connected components of a graph

In this section we start the investigation of the parallel solution of some problems linked to graphs.

The first problem we approach is the determination of the connected components of a graph $G = (V, E)$, which consists of partitioning the graph in terms of the minimum number of its connected subgraphs.

The problem of computing the connected components is considered to be one of the crucial algorithmic problems on graphs. We now present a parallel algorithm developed by Hirschberg, Chandra and Sarwate [HCS79].

The algorithm is organized in $O(\log |V|)$ phases. In each phase the vertices of the graph G are distributed on a forest of directed trees, each directed arc connecting a vertex to its father. All the vertices of the same tree belong to the same connected component of G. In the first phase of the algorithm, each vertex is the only vertex of the tree to which it belongs. At the end of the last phase all vertices belonging to the same connected component are in the same tree with height one. The transition from one phase to the next consists of the following procedure: some of the trees containing adjacent vertices of G are connected by means of a process called *hooking*, then the heights of the new trees are compressed (through the *pointer jumping* procedure treated in Section 4.3.2). The hooking process must be performed carefully, so that the tree structure is maintained and the number of phases is limited to $O(\log |V|)$. Regarding this, see the solutions proposed in [HCS79]. Concerning the PRAM implementation, it is possible to derive from the general scheme just described algorithms using time $O(\log n)$ on a CRCW PRAM with $O(n + m)$ processors, where n and m are the number of vertices and arcs in G respectively.

According to the relations among the different PRAM models (see Section 3.4), it follows that the computation of the connected components can be performed in time $O(\log n \log (n + m))$ on an EREW PRAM with $O(n + m)$ processors.

4.3.5 Sparse matrices and graphs

There are numerical linear algebra problems, such as the solution of linear systems and linear programming, in which many of the input data are zero, so that it is convenient to store them in lists instead of in vectors or matrices. It follows that these numerical problems are mapped to problems on graphs and can be solved by a combination of numeric and combinatorial techniques. These problems are called *sparse* problems.

The sparseness concept can be clarified in the case of matrices: an $n \times n$ matrix is said to be sparse when the number of non–vanishing elements is of order n, instead of n^2. In such a case, it is better to represent its non–null elements (and their positions) by a graph, instead of using a matrix with n rows and n columns.

Consider the linear system $A\mathbf{x} = \mathbf{b}$, where A is an $n \times n$ sparse matrix and its elements are stored in a list. If the solution[2] of the system is obtained by a direct method, the matrices generated by the procedure tend to be filled (this phenomenon is called *fill–in*), i.e. they are less sparse each time, so that the list size increases. This leads to the problem of determining algorithms and data structures. Unfortunately, the problem of determining the minimum fill–in is NP–complete, so that heuristics are used to limit the fill–in.

The solution of a linear system with a sparse coefficient matrix A is a four–step process:

1. row and column permutation of the matrix A;
2. symbolic factorization of A;
3. numeric factorization of A (e.g. LU factorization, or, for symmetric matrices defined positive, LL^T factorization);
4. solution of the system.

Step 1 consists of searching appropriate permutations to contain the fill–in. One of the most commonly used heuristics is the *minimum degree* method. It consists of searching the node with minimum degree in a graph associated to matrix A, and then of choosing the corresponding node of A where the elimination has to take place. After the elimination, the procedure is applied again to a submatrix of A, and so on.

In order to parallelize step 1, the availability of efficient parallel sorting algorithms, such as the minimum degree algorithm, is essential. In this respect a fast algorithm has been recently introduced [DK89].

Step 2 involves the implementation of a technique which can foresee which elements will be different from zero in the matrices factorizing A. This is a combinatorial problem, as it only involves evaluations on the matrix structure, and no numeric computation. The result is in fact the structure of factors L and U of A, which is needed to select the number of component factors to be computed (the others are null). At the same time it allows one to choose the data structure which is best suited to store the factors themselves.

One of the main elements of the parallel implementation of symbolic factorization is the concept of an *elimination tree* [DER86]. It contains information on the mutual independence of the various elimination phases, and therefore on the possible degree of parallelism.

---

[2] See [GVL83] for an introduction to fundamentals and numeric methods for linear algebra. Section 4.4 shows the parallel implementation of direct methods for solving linear systems.

The remaining two steps are numerical steps, for which there exist several efficient algorithms [DER86] for their solution. It follows that the crucial part of the solution process is in the first two steps.

The literature on this subject shows that the complexity of problems on sparse matrices is often dominated by combinatorial problems.

4.3.6 Sorting and merging

Let us now consider the problem of sorting the elements of a vector and the problem of merging two ordered sequences to obtain an ordered sequence. These two problems are crucial in the theory of algorithms and much work has been dedicated to them, both sequential[3] and parallel. Sorting is one of the most frequent activities in a computer. For example, many algorithms include sorting procedures in order to have efficient data access.

Basic operations in all sorting and/or merging algorithms include the comparison between two elements (a and b) with the possible exchange of positions, so that it can be assigned a unit cost to instructions like:

if ($a<b$) then exchange a and b

or, more simply, to tests like:

$a < b$

In the following we shall analyze algorithms considering only the steps involving comparisons, without considering the steps necessary to keep track of the results of comparison (this corresponds to evaluating algorithms in the comparison model shown in Chapter 3). For simplicity, we assume that the elements to be sorted, which belong to a linearly ordered set, are all different.

Since merging is often part of sorting algorithms, we will start with merging.

Formally, the merging problem can be defined as follows:

given two sorted sequences, $S_1 = (s_{11} s_{12} \dots s_{1n})$, with $s_{1j} < s_{1,j+1}$, $j = 1, 2, \dots, n - 1$, and $S_2 = (s_{21} s_{22} \dots s_{2m})$, with $s_{2i} < s_{2,i+1}$, $i = 1, 2, \dots, m - 1$, determine the sequence $T = (t_1, t_2, \dots, t_{m+n})$, with $t_i < t_{i+1}$, $i = 1, 2, \dots, n + m - 1$, and each t_i is an element of S_1 or S_2.

Let us describe a parallel merging algorithm: the inputs are two non-decreasing sequences $S_1 = (s_{11} s_{12} \dots s_{1n})$ and $S_2 = (s_{21} s_{22} \dots s_{2m})$ with length

---

[3] As far as sequential algorithms are concerned, many efficient schemata have been developed (e.g. quicksort and mergesort).

n and m ($n \leq m$). Moreover, let us assume, for the sake of simplicity, that n and m are perfect squares.[4]

Merging algorithm

1. Divide S_1 in $\sqrt{n}$ blocks of length $\sqrt{n}$ and S_2 in $\sqrt{m}$ blocks of length $\sqrt{m}$.
2. Let α_i be the first element of the ith block of S_1 and β_j the first element of the jth block of S_2; compare in parallel α_i and β_j, $i = 1, 2, ..., \sqrt{n}, j = 1, 2, ..., \sqrt{m}$.
3. Determine in parallel, for each α_i, the index $j(i)$ such that $\beta_{j(i)} < \alpha_i < \beta_{j(i)+1}$.
4. Compare α_i with all the elements of the block starting with element $\beta_{j(i)}$.

At this point the algorithm has determined the correct position of each α_i inside blocks of S_2. Hence the problem is now to solve a set of disjoint merging problems, each one consisting of merging blocks of S_1 with some subsequences of S_2. It is possible to proceed recursively.

We can evaluate the number $T(n)$ of steps of the algorithm, keeping in mind that the computation model considers only comparisons.

It is easy to verify that $T(n)$ consists of two comparisons (phase 2 and 4) to which one must add the solution of merging problems of order $\sqrt{n}$, so that:

$$T(n) \leq T(\sqrt{n}) + 2$$

from which we obtain:

$$T(n) = O(\log\log n)$$

Borodin and Hopcroft [BH82] have shown that this algorithm can be implemented on a CREW PRAM with $O(n + m)$ processors in time $O(\log\log n)$, which is optimal.

Let us now consider the sorting problem. Formally, the problem can be defined as follows:

Given a sequence S of elements s_i, $i = 1, ..., n$, on which a linear ordering is defined, determine the sequence $T = (t_1, t_2, ..., t_n)$, such that $t_i < t_{i+1}$ and $t_i = s_{p(i)}$, where p is a permutation of $\{1, 2, ..., n\}$.

---

[4]This assumption could seem too heavy. Actually, the general case can be treated with trivial changes, but we do not consider them for the sake of simplicity.

One of the classic parallel sorting algorithms is the *bitonic* sorting introduced by Batcher. The approach is based on some properties of sequences of elements.

Let $S = (s_1 s_2 ... s_n)$ be a sequence of distinct elements belonging to a set where some ordering criteria is defined:

- An element s_i is called a *local minimum* if both s_{i-1} and s_{i+1} are greater than s_i, $i = 2, 3, ..., n - 1$;
- an element s_i is called a *local maximum* if both s_{i-1} and s_{i+1} are smaller than s_i, $i = 2, 3, ..., n - 1$;
- a sequence S is called *unimodal* if it contains at most a local minimum or maximum;
- a sequence S is called *bitonic* if it consists of a cyclic shift of a unimodal sequence.

Bitonic sequences have the following properties.

Let $S = (s_1 s_2 ... s_{2n})$ be a bitonic sequence. Let us define the sequences $m(S)$ and $M(S)$ as follows:

$m(S) = (\min\{s_1, s_{n+1}\}, \min\{s_2, s_{n+2}\}, ..., \min\{s_n, s_{2n}\})$
$M(S) = (\max\{s_1, s_{n+1}\}, \max\{s_2, s_{n+2}\}, ..., \max\{s_n, s_{2n}\})$

We have that:

- both $m(S)$ and $M(S)$ are bitonic sequences;
- each element of $m(S)$ is smaller than the corresponding element of $M(S)$.

The above property suggests the following parallel algorithm for sorting bitonic sequences, which is based on the *divide et impera* technique.

Algorithm for sorting bitonic sequences

Let $n = 2^k$;
SORT(S)
begin
    if (S has length 1)
    then stop
    else {make $m(S)$ and $M(S)$;
        SORT($m(S)$) and SORT($M(S)$) in parallel}
end.

The procedure, applied to the bitonic sequence (1, 2, 3, 8, 7, 6, 5, 4), is shown hereafter:

```
            1 2 3 8 7 6 5 4
              /         \
             /           \
            /             \
         1 2 3 4        7 6 5 8
          /  \           /  \
         /    \         /    \
        1 2   3 4     5 6    7 8
```

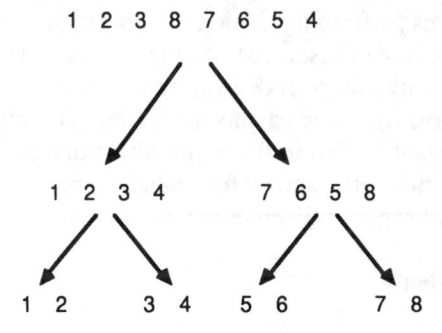

The parallel cost of the algorithm is $T = O(\log n)$, if $n/2$ computing units are available.

Let us now present an algorithm (due to Batcher) for sorting an arbitrary sequence, which is based on the previous one.

Batcher sorting algorithm (bitonic sorting)

Let $n = 2^k$. An arbitrary sequence S of length n is considered as $n/2$ bitonic sequences of length 2. Starting from these sequences we build $n/4$ bitonic sequences of length 4, and then $n/8$ bitonic sequences of length 8, and so on, until we obtain a single bitonic sequence of length n. At this point the algorithm for bitonic sequences can be applied.

The algorithm is based on the fact that the concatenation of an increasing sequence with a decreasing one produces a bitonic sequence. It is then possible to convert $n/2^i$ bitonic sequences of length 2^i into $n/2^{i+1}$ bitonic sequences of length 2^{i+1}. This is done by sorting the sequence of length 2^i in increasing and decreasing order alternatively, using the previous algorithm for sorting bitonic sequences.

Let us evaluate the performance of the algorithm. The algorithm for bitonic sequences is used to form bitonic sequences of length 2^{i+1} starting from sequences of length 2^i. Since this transformation takes time $O(i)$, it follows that:

$$T(n) = T(2^k) = O(\sum_{i=1}^{k} i) = O(\log^2 n)$$

The maximum number of comparisons to be executed in parallel is $n/2$, so that the amount of hardware sufficient to obtain the maximum parallelism is $n/2$.

The bitonic sorting algorithm can be implemented by an EREW PRAM with $n/2$ processors. The computation time remains $O(\log^2 n)$.

The comparisons performed are of the order of $n\log^2 n$. Since there exist sequential algorithms performing $O(n\log n)$ comparisons, the bitonic sorting involves a redundant use of resources (i.e. the efficiency is of order $1/\log n$).

In 1983 Ajtai and others [AKS83] developed a parallel algorithm performing $O(n\log n)$ comparisons in $O(\log n)$ time. However, the multiplicative constant hidden in the asymptotic notation is very large, and this makes the algorithm unpractical for realistic input sizes. For this reason the bitonic algorithm is generally preferred.

Cole sorting algorithm

In 1986 Cole presented a sorting algorithm reaching the same asymptotic performance of the Ajitai algorithm on EREW PRAMs and sufficiently practical to be used in real implementations. The rest of this section sketches this algorithm, without providing the details necessary to prove its correctness. Let n be the number of elements (all distinct) to be sorted. For simplicity we assume that n is a power of 2.

The $n = 2^k$ elements can be viewed as arranged on the leaves of a complete binary tree A with depth k. Given an internal node v, A_v indicates the subtree of A rooted at v. To each node is associated an ordered list L_v, formed by the leaves of A_v (for simplicity we do not distinguish leaves from the corresponding elements). At the beginning of step s, the list associated to node v is indicated with $L_v(s)$. Initially, all lists are empty, except those associated to leaves. A node v is called *border* if L_v contains all the leaves of A_v and if its immediate predecessor is not a border. Each node predecessor of a border is said to be *internal*. As the algorithm proceeds, the border moves from the leaves towards the root. The algorithm ends when the only border node is the root. Let us now consider, for each internal or border node v, the list $U_v(s+1)$ defined in the following way:

- If v is an internal node, or if s is the first step in which v is a border, $U_v(s+1)$ is formed with the elements in position 1, 5, 9, ... in the list $L_v(s)$.
- If s is the second step in which v is a border, $U_v(s+1)$ is formed with the elements in position 1, 3, 5, ... in the list $L_v(s)$.
- If s is the third step in which v is a border, $U_v(s+1)$ contains all the elements of $L_v(s)$.

The generic sth step of the algorithm includes the following actions, executed in correspondence of each internal node v.

1. Make the lists $U_x(s+1)$ and $U_y(s+1)$, where x and y are the immediate successors of v.
2. Compute the new list $L_v(s+1)$ by merging $U_x(s+1)$ and $U_y(s+1)$.

It is easy to realize that if a node becomes a border at step s, its immediate predecessor becomes a border at step $s + 3$. The sorting procedure ends in $3k = 3\log n$ steps. The cost of each step depends on the time needed to merge the lists. Cole proved that this operation can be performed in constant time using a processor for each element to be sorted. Note that the sum of the lengths of all lists, L and U, associated to internal or border nodes, is, at any given instant, $O(n)$. The details of the proof, which is rather long and articulated, have been omitted for the sake of simplicity.

4.4 Numerical algorithms

This section faces some of the crucial problems in numerical linear algebra: the solution of linear systems, the computation of the inverse, the determinant and the rank of a matrix. Foundations and numerical methods for linear algebra are treated in the books [FF63], [Ho64], [GVL83] and [BCM88]. According to the model of Section 3.2, the cost of algorithms will hereafter be expressed in terms of the pair $PT = (P(n), T(n))$, indicating the hardware (processor number) and the time, as a function of the size of the problem.

Parallelism in linear algebra problems has been studied since the sixties, with a major increase in the seventies. Munro assumed that it was not possible to compute the inverse of an $n \times n$ matrix using a sublinear number of steps. This hypothesis was later proved false. In fact, in 1976, Csanky [Cs76] presented an algorithm for linear system solution (and matrix inversion) running in $O(log^2 n)$ time on $O(n^4)$ processors. Csanky's algorithm is only interesting from a theoretical point of view, since it requires too many processors and is numerically unstable.

4.4.1 Elimination methods

One of the first attempts to develop parallel algorithms for solving linear systems is due to Pease, who in 1967 analyzed the *Gauss* and the *bordering* methods. The parallel version of these two algorithms, as developed by Pease, has a cost $PT(n, n^2)$ [Pe67]. Pease's work is based on a search for efficient parallel algorithms starting from sequential ones (not necessarily efficient).

In the context of methods originated from efficient sequential algorithms, let us now describe the parallel version of the Gauss–Jordan and Givens factorization methods.

The Gauss–Jordan algorithm solving a system of linear equations can be implemented in parallel with cost $PT(n^2, n)$. In fact, if we consider the execution of the kth step, the "modified" coefficient matrix $A^{(k)}$ and the modified known term $b^{(k)}$ have the following structure:

150 Parallel algorithms

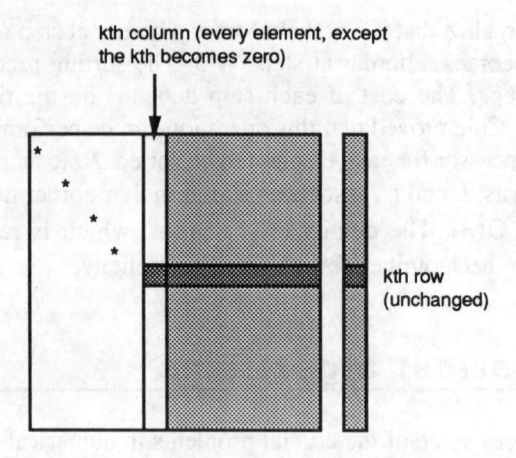

kth column (every element, except the kth becomes zero)

kth row (unchanged)

Assume that $a_{ij}^{(k)}$ and $b_i^{(k)}$, $1 \leq i, j \leq n$, denote the element (i, j) of $A^{(k)}$ and the element i of $b^{(k)}$ respectively. The kth step of the Gauss–Jordan elimination is given by:

$$a_{ij}^{(k+1)} = a_{ij}^{(k)} - \frac{a_{ik}^{(k)}}{a_{kk}^{(k)}} a_{kj}^{(k)},$$

$$i = 1, \ldots, k-1, k+1, \ldots, n$$
$$j = k+1, \ldots, n$$

$$b_i^{(k+1)} = b_i^{(k)} - \frac{a_{ik}^{(k)}}{a_{kk}^{(k)}} b_{kj}^{(k)},$$

The element $a_{kk}^{(k)}$ is called the *pivot*.

It is easy to verify that, when the pivot is known, a whole submatrix can be eliminated in one parallel step. Hence the bound $PT(n^2, n)$ anticipated above.

Similarly to the sequential case, the parallel version of the Gauss–Jordan method fails if $a_{kk}^{(k)} = 0$, for any k. In order to avoid this degeneracy, we can search a (partial) pivot on the columns of $A^{(k)}$. However, this introduces additional costs, due to the algorithm searching the maximum among the elements $|a_{kk}^{(k)}|, \ldots, |a_{nk}^{(k)}|$, which involves additional $\lceil \log(n - k + 1) \rceil$ steps. The total cost of the algorithm with partial pivot is then $PT(n^2, n\log n)$. It follows that the asymptotic cost of the Gauss–Jordan method with partial pivot is, contrary to what happens in the sequential case, greater than the cost of the Gauss–Jordan method.

The loss of performance due to the pivot search led to the parallelization of some orthogonalization methods, such as Householder and Givens (QR). As an example, Sameh and Kuck [SK78] introduced a technique to parallelize the computation of the Givens rotation, which allowed the performance $PT(n^2, n)$ to be obtained. In order to understand the approach adopted by Sameh and Kuck, consider the (sequential) Givens procedure:

$$QA = R$$
$$Q = Q_{n-1} \ldots Q_2 Q_1,$$
$$Q_j = P_{j,j+1}^{(j)} \ldots P_{n-1,n}^{(j)}, \quad j = 1, \ldots, n-1$$

where the matrices $P_{i,i+1}^{(j)}$ are special *orthogonal* matrices (see Figure 4.8), each of which is obtained by adding a rank 2 matrix to the identity matrix [BCM88].

The value of ϕ can be chosen in such a way that the element in position $(i + 1, j)$ of the matrix:

$$P_{i,i+1}^{(j)} A_i^{(j)}$$

where

$$A_i^{(j)} = P_{i-1,i}^{(j)} \ldots P_{n-1,n}^{(j)} Q_{j-1} \ldots Q_2 Q_1 A$$

can be eliminated maintaining the zeros previously introduced. Since each product modifies only the jth and the $(i + 1)$th rows of $A_i^{(j)}$, it is possible to

$$P_{i,i+1}^{(j)} = \begin{bmatrix} 1 & & & & & & & & \\ & \cdot & & & & & & & \\ & & 1 & & & & & & \\ & & & c & \cdot & \cdot & \cdot & -s & \\ & & & \cdot & 1 & \cdot & \cdot & \cdot & \\ & & & \cdot & \cdot & \cdot & \cdot & \cdot & \\ & & & \cdot & \cdot & \cdot & 1 & \cdot & \\ & & & s & \cdot & \cdot & \cdot & c & \\ & & & & & & & & 1 \\ & & & & & & & & & \cdot \\ & & & & & & & & & & 1 \end{bmatrix}$$

Figure 4.8 Givens elementary matrix ($s = \sin \phi$ and $c = \cos \phi$).

152 Parallel algorithms

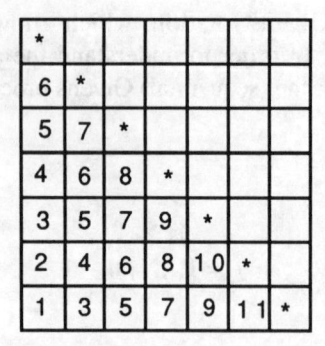

*						
6	*					
5	7	*				
4	6	8	*			
3	5	7	9	*		
2	4	6	8	10	*	
1	3	5	7	9	11	*

Figure 4.9 Sameh and Kuck eliminating scheme.

conceive an elimination method in which the original matrix is multiplied each time by the direct sum of more than one Givens matrix. Sameh and Kuck developed a similar scheme, which can be implemented in $2n - 3$ steps. The process eliminates the elements of A according to the ordering shown in Figure 4.9 (see [SK78]), where the positions indicated with k are considered at the kth step.

Note that the methods presented until now lead in the best case to linear time algorithms. Simple arguments on problem data, on the dependence of the result from each input, and on the bounds on operation arity, prove that a lower bound on the number of steps for solving either an $n \times n$ linear system, or the inversion of an $n \times n$ matrix, is $2\log n$. It follows that the elimination methods, even though they reach good performances in terms of speed-up and efficiency, leave a significant gap with respect to the time lower bound.

4.4.2 Csanky algorithm

Csanky's result has considerably reduced the gap between the best available algorithm and the lower bound $2\log n$. The upper bound corresponding to Csanky's algorithm ($T = O(\log^2 n)$) does not eliminate the gap, but it is a remarkable improvement. The contribution given by Csanky is important also because it shows that sometimes it is possible to obtain extremely fast parallel algorithms only if we abandon the idea of developing parallel algorithms by transforming sequential algorithms.

Csanky obtained an efficient parallel algorithm for matrix inversion starting from some results of linear algebra and parallel computing:

1. the recursion relations, known as the Newton identity [FF63], $kc_k = -s_k - c_1 s_{k-1} - \ldots - c_{k-1} s_1$, $k = 1, 2, \ldots, n$, where $s_i = \text{tr}(A^i)$, $i = 1, 2, \ldots, n$, which are used (in Leverrier method) to determine the coefficients $c_1, c_2, \ldots, c_n$ of the characteristic polynomial of matrix A:

$$\begin{bmatrix} 1 & & & & & \\ s_1 & 2 & & & & \\ s_2 & s_1 & 3 & & & \\ \cdot & \cdot & \cdot & & & \\ \cdot & \cdot & \cdot & \cdot & & \\ \cdot & \cdot & \cdot & \cdot & \cdot & \\ s_{n-1} & s_{n-2} & \cdot & \cdot & s_1 & n \end{bmatrix} \begin{bmatrix} c_1 \\ c_2 \\ c_3 \\ \cdot \\ \cdot \\ \cdot \\ c_n \end{bmatrix} = - \begin{bmatrix} s_1 \\ s_2 \\ s_3 \\ \cdot \\ \cdot \\ \cdot \\ s_n \end{bmatrix} \qquad (1)$$

2. one of the existing algorithms for inverting an $n \times n$ triangular matrix in time $T = O(\log^2 n)$;
3. the Cayley–Hamilton theorem [FF63], stating that a matrix is a root of its characteristic polynomial, i.e. $A^n + c_1 A^{n-1} + \ldots + c_n I = 0$, from which it immediately follows that, if $c_n \neq 0$:

$$A^{-1} = - \frac{A^{n-1} + c_1 A^{n-2} + \ldots + c_{n-1} I}{c_n} \qquad (2)$$

The problem is then reduced to determining the powers $A^2, A^3, \ldots, A^n$. Using the classic fan–in algorithm to multiply $n \times n$ matrices in $PT(n^3, \log n)$, the matrix powers can be computed in $PT(n^4, \log^2 n)$ by an algorithm with the following structure:

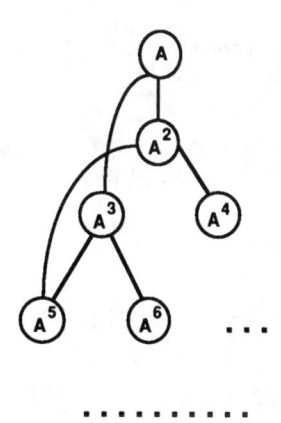

Note that the solution of system (1) can also be obtained by computing the inverse of the coefficient matrix S using formula (2):

$$S^{-1} = -\frac{S^{n-1} + d_1 S^{n-2} + \ldots + d_{n-1} I}{d_n} \qquad (3)$$

where the coefficients d_i of the characteristic polynomial of S can be computed in time $O(\log n)$ starting from the zeros (i.e. the diagonal elements $1, 2, \ldots, n$ of S). Hence if $I(n)$ is the depth of an arithmetic circuit computing the first n powers of an $n \times n$ matrix, then:

$$I(n) \leq 2\text{PWR}(n) + O(\log n)$$

Before describing the Csanky algorithm, let us briefly consider three methods [SB77] which can be used for solving the triangular system occurring in the algorithm. These methods provide formulas for the inverse of a triangular matrix which can be computed in time $O(\log^2 n)$.

Let $L_n = (l_{ij})$ be an $n \times n$ lower triangular matrix, with $l_{ii} \neq 0$, $i = 1, 2, \ldots, n$.

Method a

Let us assume, without loss of generality, that $l_{ii} = 1$. L_n can then be rewritten as $L_n = I - H$, where $H^n = 0$. As the spectral radius of the matrix H is less than one (in particular it is equal to zero), then L_n^{-1} has the expression $L_n^{-1} = I + H + H^2 + \ldots + H^{n-1} + H^n + \ldots$, from which it follows that

$$L_n^{-1} = I + H + H^2 + \ldots + H^{n-1} \qquad (4)$$

since $H^i = 0$, $i \geq n$.

Method b

L_n can be written as:

$$L_n = \begin{bmatrix} L'_{n/2} & 0 \\ B_{n/2} & L''_{n/2} \end{bmatrix}$$

where $L'_{n/2}$ and $L''_{n/2}$ are $(n/2) \times (n/2)$ triangular matrices and $B_{n/2}$ is an $(n/2) \times (n/2)$ matrix. Then:

$$L_n^{-1} = \begin{bmatrix} L'^{-1}_{n/2} & 0 \\ -L''^{-1}_{n/2} B^{-1}_{n/2} L'^{-1}_{n/2} & L''^{-1}_{n/2} \end{bmatrix} \qquad (5)$$

which suggests a recursive algorithm for computing L_n^{-1} in time $T(n)$, where:

$$T(n) \leq T(n/2) + O(\log n)$$

i.e. $T(n) = O(\log^2 n)$.

Method c

Let us assume, without loss of generality, that $l_{ii} = 1$. Let v_i, $i = 1, 2, ..., n$, be the ith column of matrix $H = L_n - I$, and let e_i be the ith column of the identity matrix. L_n can then be written as:

$$L_n = \prod_{i=1}^{n-1} (I + v_i e_i^T)$$

from which we obtain:

$$L_n^{-1} = \prod_{i=n-1}^{1} \left(I + \frac{1}{1 + e_i^T v} v_i e_i^T\right) \qquad (6)$$

It is easy to verify that expressions (4), (5) and (6) originate algorithms for inverting (or for solving the associated system) matrix L_n with a parallel cost $PT(n^3, \log^2 n)$.

We can now summarize the Csanky algorithm:

1. compute the powers $A^2, A^3, ..., A^n$ and their traces;
2. form the triangular matrix S;
3. compute the powers $S^2, S^3, ..., S^n$;
4. compute the coefficients $d_1, ..., d_n$ of the characteristic polynomial of S, whose roots are $1, ..., n$;
5. compute S^{-1} using formula (3);
6. compute the coefficients $c_1, ..., c_n$ of the characteristic polynomial of A through $[c_1, ..., c_n]^T = S^{-1}[s_1, ..., s_n]^T$;
7. compute A^{-1} through formula (2).

Note that steps 3–5 can be replaced by the computations of the inverse of S through one of the methods (4), (5) or (6).

4.4.3 Improvements to the Csanky algorithm

The main negative aspects of the Csanky algorithm are its numerical instability due to the computation of the matrix powers, and the high demand of hardware, $P = O(n^4)$. The request for hardware is strictly linked to that due to multiplying two matrices. For the latter problem, the value n^3 can be easily reduced to $n^3/\log n$, by using the technique shown in Section 4.1.5. Additional reductions based on fast sequential algorithms for the multiplication of matrices, such as the Strassen algorithm [St69], lead to performances $PT(n^\alpha/\log n, \log n)$, $2 < \alpha < 3$.

Preparata and Sarwate [PS78] used these results to develop an algorithm computing the inverse of a real matrix in $PT(2n^{\alpha+(1/2)}/\log^2 n, \log^2 n)$, executing not more than $n^{1/2}$ matrix multiplications. Their improvement with respect to Csanky is due to two reasons:

- only the diagonal elements of the matrices $A^2, ..., A^n$ are necessary to implement the Leverrier method;
- it is possible to compute the matrix $A^{n-1} + c_1 A^{n-2} + ... + c_{n-1} I$ by means of a recurrence relation involving at most $n^{1/2}$ matrix multiplications at each phase, always maintaining the $O(\log^2 n)$ time.

For these computations, roughly $2n^{1/2}$ powers of matrix A are required, i.e. $A, A^2, ..., A^p, A^{2p}, ..., A^{p^2}$, where $p = \lceil n^{1/2} \rceil$. The algorithm of Preparata and Sarwate follows.

1. Compute the powers $A^2, ..., A^p$. This can be done in $PT(pn^\alpha/((\log n)(\log p)), (\log n)(\log p) = PT(2n^{\alpha+(1/2)}/\log^2 n, \log^2 n)$, where the additional improvement of a factor $\log p$ in the hardware is due to the organization of the algorithm.
2. Using the same subcircuit, compute the powers $B^2, ..., B^p$, where $B = A^p$.
3. Let $A_{j*}^{(i)}$ and $A_{*j}^{(i)}$, $j = 1, ..., n$, be the jth row and the jth column of matrix A^i. Form n $p \times n$ matrices R_j, and n $n \times p$ matrices T_j, where:

$$R_j = \begin{bmatrix} A_{j*}^{(0)} \\ A_{j*}^{(1)} \\ \vdots \\ A_{j*}^{(p-1)} \end{bmatrix} \quad \text{and} \quad T_j = \begin{bmatrix} A_{*j}^{(p)} & A_{*j}^{(2p)} & \cdots & A_{*j}^{(p^2)} \end{bmatrix}$$

This can be done with $P \leq n^{\alpha+(1/2)}$.

4. Compute the n products $U_j = R_j T_j$. U_j elements are the elements (j, j) of the matrices A^k, $k = p, p+1, \ldots, p^2 + p - 1$. In this way, simply by multiplying $p \times n$ matrices by $n \times p$ matrices, it is possible to compute the traces of all the powers. Each product corresponds to p products of two $p \times p$ matrices, so that all products can be computed with $PT(npp^\alpha/\log p, \log n) = PT(2n^{(1/2)(3+\alpha)}/\log n, \log n)$.

5. Using the coefficients computed in 4 and the matrices $I, A, \ldots, A^n$, form the p matrices:

$$B_0 = c_{n-1} I + c_{n-2} A + \ldots + c_{n-p} A^{p-1}$$

$$B_1 = c_{n-p-1} I + c_{n-p-2} A + \ldots + c_{n-2p} A^{p-1}$$

$$B_{p-1} = c_{n-p^2+p-1} I + c_{n-p^2+p-1} A + \ldots + c_{n-p^2} A^{p-1}$$

where $c_0 = 1$ and $c_k = 0$ if $k < 0$. Note that the jth row of B_{i-1} is the ith row of the matrix CR_j, where:

$$C = \begin{bmatrix} c_{n-1} & c_{n-2} & \cdots & c_{n-p} \\ c_{n-p-1} & c_{n-p-2} & \cdots & c_{n-2p} \\ \cdots & \cdots & \cdots & \cdots \\ c_{n-p^2+p-1} & c_{n-p^2+p-2} & \cdots & c_{n-p^2} \end{bmatrix}$$

It is then possible to compute the matrices B_i in $PT(npp^\alpha/\log p, \log n) = PT(2n^{(1/2)(3+\alpha)}/\log n, \log n)$.

6. Let $h = \lfloor \log p \rfloor$ and $l = \lceil p/h \rceil$. Compute the recurrence relation:

$$D_i = D_{i-1} + \sum_{r=(i-1)l}^{il-1} B_r A^{pr}, \quad i = 1, \ldots, h$$

$$D_{h+1} = D_h + \sum_{r=lh}^{p-1} B_r A^{pr}$$

This involves O(log p) phases, each one consisting of O(p/log p) matrix multiplications and additions, so that $PT((p/\log p)n^\alpha/\log n, (\log p)(2\log n)) = PT(2n^{\alpha+(1/2)}/\log^2 n, \log^2 n)$.

7. Compute $A^{-1} = -D_{h+1}/c_n$.

Since $\alpha > 2$, the total parallel cost is $PT(2n^{\alpha+(1/2)}/\log^2 n, \log^2 n)$.

The algorithm of Preparata and Sarwate gives a substantial improvement in the hardware complexity with respect to the Csanky algorithm. However, this improvement is obtained at the expense of simplicity.

4.4.4 Extensions to the Csanky algorithm

We have seen that the methods based upon the Cayley–Hamilton theorem can be used to solve some linear algebra problems. In every case these methods require the execution of a division, i.e. $1/d_n$, where d_n is the coefficient of the term with zero degree of the characteristic polynomial of matrix S. Hence these methods can be applied only to those problems which are defined on fields containing integers. Moreover, as $d_n = n!$, they do not seem to be applicable to the case of finite fields. It is however very important to solve systems with coefficient matrices having elements belonging to general algebraic structures, such as the rings with unity. In 1984, Berkowitz [Be84] showed that the characteristic polynomial, the determinant and the adjoint of an $n \times n$ matrix A, with elements belonging to an arbitrary ring, can be computed with $T = O(\log^2 n)$ using only the adding constants -1, 0 and 1. The main result, allowing one to avoid *non-closed* operations, is the Samuelson method. It relates the characteristic polynomial of A to the characteristic polynomial of its submatrix M of order $n-1$, where:

$$A = \begin{bmatrix} a_{11} & R^t \\ Z & M \end{bmatrix}$$

and where R^t, Z and M have size $1 \times (n-1)$, $(n-1) \times 1$ and $(n-1) \times (n-1)$ respectively. Samuelson states that:

$$p(\lambda) = (a_{11} - \lambda)\det(M - \lambda I) - R^t(\sum_{k=2}^{n}(q_0 M^{k-2} + \ldots + q_{k-2} I)\lambda^{n-k})Z$$

or, in matrix form:

Numerical algorithms

$$\begin{bmatrix} p_0 \\ p_1 \\ \vdots \\ p_n \end{bmatrix} = \begin{bmatrix} -1 & & & & & \\ a_{11} & -1 & & & & \\ -R^t Z & a_{11} & -1 & & & \\ -R^t MZ & -R^t Z & a_{11} & -1 & & \\ \cdots & \cdots & \cdots & \cdots & \cdots & \\ -R^t M^{n-4} Z & -R^t M^{n-5} Z & \cdots & & a_{11} & -1 \\ -R^t M^{n-3} Z & -R^t M^{n-3} Z & \cdots & & -R^t Z & a_{11} \end{bmatrix} \begin{bmatrix} q_0 \\ q_1 \\ \vdots \\ q_{n-1} \end{bmatrix}$$

where the coefficient matrix C is an $n \times (n-1)$ Toeplitz[5] matrix. By recursively applying the formula to matrix M it follows that $p(\lambda)$ is the product of n Toeplitz matrices $C = C_0, C_1, \ldots, C_{n-1}$, where C_i has dimension $(n-i) \times (n-i-1)$ (assuming that a 1×0 matrix is a scalar):

$$(p_0 \, p_1 \cdots p_n)^t = \prod_{i=0}^{n-1} C_{n-i}$$

Hence the computation of the characteristic polynomial of A can only be performed by means of additions and multiplications. Since two $n \times n$ Toeplitz matrices can be multiplied in $PT(n, \log n)$, the cost of evaluating the above formula using a balanced binary tree is $PT(n^2, \log^2 n)$. In any case, it is first necessary to compute the elements of the Toeplitz matrices. This involves computing the following values:

$$H_m = \left\{ R_m^t M_m^k Z_m \right\}_{k=0}^{m}, \quad m = 1, \ldots, n-1$$

where R_m^t, M_m and Z_m have order $1 \times m$, $m \times m$, and $m \times 1$, respectively. Berkowitz showed that this computation can be performed in $PT(n^{1+\alpha+\varepsilon}, \log^2 n)$, with $\varepsilon > 0$. Hence the parallel cost of computing the characteristic polynomial is $PT(n^{1+\alpha+\varepsilon}, \log^2 n)$. The proof given by Berkowitz is given below, where the notation n^x is used to denote $\lceil n^x \rceil$.

Theorem 4.3. Let R^t be a row vector of order m, and let M be an $m \times m$ matrix, with $m < n$. The set of vectors:

$$X_\beta = \left\{ R^t M^i \right\}_{i=0}^{n^\beta}$$

---

[5] A matrix $T = (t_{ij})$ is a Toeplitz matrix if $t_{ij} = t_{i-j}$, i.e. if the elements on the same diagonal are equal.

with $\beta < 1$, can be computed in $PT(n^{\alpha+\varepsilon}, \log^2 n)$.

Proof. By induction on the number of elements of X_β. The thesis is trivially true for $X_{-\infty} = \{R^t\}$. Let us now assume the hypothesis is true for $X_{\beta-\varepsilon}$ and let us consider the matrix:

$$W_{\beta-\varepsilon} = \begin{bmatrix} R^t \\ R^t M \\ \vdots \\ R^t M^{n^{\beta-\varepsilon}} \end{bmatrix}$$

the rows of which are the vectors of $X_{\beta-\varepsilon}$. Let:

$$Y_\beta = \left\{ I, M^{n^\beta}, M^{2n^\beta}, \ldots, M^{n^{\beta+\varepsilon}} \right\}$$

Multiplying $W_{\beta-\varepsilon}$ by the matrices of Y_β we obtain all the vectors of X_β. This can be done in $PT(n^{\alpha+\varepsilon}, \log^2 n)$ (n^ε multiplications of $(n^{\beta-\varepsilon} \times n) \times (n \times n)$ matrices, with $\beta-\varepsilon < 1$). Moreover the computation of the set Y_β can be performed in $PT(n^{\alpha+\varepsilon}, \log^2 n)$.

The previous theorem applies also to the problem of computing the set

$$U_\beta = \left\{ M^{in^\beta} Z \right\}_{i=0}^{n^\beta}$$

Proposition 4.4
The sets H_m, $m = 1, \ldots, n-1$, can be computed in $PT(n^{1+\alpha+\varepsilon}, \log^2 n)$.

Proof. For $m = 1, \ldots, n-1$, compute the sets:

$$X_{m, 0.5} = \left\{ R_m^t M_m^i \right\}_{i=0}^{n^{0.5}} \quad \text{and} \quad U_{m, 0.5} = \left\{ M_m^{jn^{0.5}} Z_m \right\}_{j=0}^{n^{0.5}}$$

This can be done in $PT(n^{1+\alpha+\varepsilon}, \log^2 n)$ as follows from the previous theorem. Each element of H_m, $m = 1, \ldots, n-1$, can be computed as a scalar product of vectors starting from $X_{m,0.5}$ and $U_{m,0.5}$. In fact, all exponents of terms like M_m present in H_m can be expressed as $i + jn^{0.5}, 0 \le i, j \le n^{0.5}$. Since all scalar products can be computed in $PT(n^3, \log n)$, the total parallel cost is $PT(n^{1+\alpha+\varepsilon}, \log^2 n)$.

It is possible to obtain the determinant of a matrix A provided that its characteristic polynomial $p(\lambda)$ is available, by simply letting λ be zero in $p(\lambda)$. The sum matrix can then be computed as:

$$\text{adj}(A) = A^{n-1} + p_1 A^{n-2} + \ldots + p_{n-1} I$$

Finally, it is important to remember a result for the computation of the rank of a matrix with elements in any field [Mu86]. The result leads to an algorithm with performance $PT(n^{4.5}, \log^2 n)$. It is based on reducing the problem of computing the rank of a matrix A to the problem of computing the characteristic polynomial of the matrix:

$$\begin{pmatrix} 0 & A^T \\ A & 0 \end{pmatrix}$$

4.4.5 Iterative methods

This section considers two iterative methods: the Newton method and asynchronous iterations. Iterative methods differ from direct methods (to this class belong all the methods illustrated until now) because, even if an arithmetic with an infinite number of digits were used, it would not compute the exact solution.

Let F be an operator from a space S into itself. An *iterative method* corresponding to the operator F is a sequence:

$$s^{(0)}, s^{(1)}, \ldots, s^{(k)}, \ldots$$

of points in S, in which $s^{(k+1)} = F(s^{(k)})$, and $s^{(0)}$ is chosen arbitrarily. We say that an iterative method converges to a fixed point ξ of F if:

$$\lim_{i \to \infty} s^{(i)} = \xi < +\infty$$

The notation $\xi < +\infty$ means that a norm of the vector (or matrix) ξ is finite.

Parallel implementation of the Newton method

Let us apply the Newton method to the computation of the inverse of a real $n \times n$ matrix A. We can start from the equation:

$$AX - I = 0$$

which leads to the iterative method:

$$\begin{cases} X_{k+1} = X_k - X_k R_k \\ R_{k+1} = A X_k - I \end{cases} \quad k = 0, 1, \ldots$$

or, noting that $R_{k+1} = -R_k^2$ and setting $P_k = -R_k$, to:

$$\begin{cases} X_{k+1} = X_k + X_k P_k \\ P_{k+1} = P_k^2 \end{cases} \quad k = 0, 1, \ldots$$

Each step of the last iterations consists of two matrix multiplications and one matrix addition. Hence the parallel cost of each step is $PT(n^\alpha/\log n, \log n)$. An analysis of the total cost has been done in [PR85], where it has been proven that the number of steps is of the order of $\log(\text{Cond}(A))$, and the total cost is:

$PT(n^\alpha/\log n, \log n \log(\text{Cond}(A)))$

It follows that the Newton method is comparable with the Csanky method and its variations, subject to the condition that the coefficient matrix is *well conditioned*.[6]

Iterative asynchronous methods

A completely different class of iterative methods is the class of asynchronous or chaotic iterations [CM69], [Ba78].

We will see in Chapter 5 that the main advantage of asynchronous computation is that the computing elements do no have to synchronize. Each

---

[6]In the context of asymptotic analysis, an $n \times n$ matrix A is said to be well conditioned if $\text{Cond}(A)$ does not increase too quickly with n. A reasonable criterion is to distinguish between polynomial and exponential conditioning.

processor is then free to run at its own speed. Let us formally discuss asynchronous iterations.

Let F be a linear operator from R^n to R^n. F can be represented as:

$$F(\mathbf{x}) = A\mathbf{x} + \mathbf{b}$$

Starting from vector $\mathbf{x}^{(0)}$, a synchronous iterative method is a sequence:

$$\mathbf{x}^{(0)}, \mathbf{x}^{(1)}, \ldots$$

of vectors of R^n defined as follows.

Let $J = J_1, J_2, \ldots$ be a sequence of non–empty subsets of $\{1, 2, \ldots, n\}$, and $S = S_1, S_2, \ldots$ be a sequence of elements of N^n. Each S_j is defined by means of the vectors $[s_1(j), s_2(j), \ldots, s_n(j)]$, where the functions $s_i: N \to N$, $i = 1, \ldots, n$, are subject to the following constraints:

1. $s_i(j) \leq j - 1$, for every i and j;
2. $s_i(j) \to +\infty$ for $j \to +\infty$.

Moreover, the sequence J must satisfy the following:

3. every index i, $i = 1, \ldots, n$, appears an infinite number of times in the sequence.

The ith component $x_i^{(j)}$ of the jth approximation $\mathbf{x}^{(j)}$ generated from an iterative asynchronous method can be defined as:

$$x_i^{(j)} = \begin{cases} x_i^{(j-1)} & \text{if } i \notin J_j \\ f_i(x_1^{(s_1(j))}, \ldots, x_n^{(s_n(j))}) & \text{if } i \in J_j \end{cases}$$

Constraints 1–3 allow us to adopt several different schemas. The most natural criterion is to take the most recently updated components.

As far as the convergence to the fixed point is concerned, it is sufficient that the function F is a (Lipschitz) contraction[7] from a closed subset D of R^n in $F(D)$ (and $D \supset F(D)$). In this case, for each initial vector $\mathbf{x}^{(0)}$, the method converges to the only fixed point ξ of F in D. For linear operators which are contractions:

$$F(\mathbf{x}) = A\mathbf{x} + \mathbf{b}$$

---

[7] A function $f: R^n \to R^n$ is a contraction if $\|f(x) - f(y)\| < \|x - y\|$, for every $x, y \in R^n$.

the matrix A satisfies the condition $\rho(A) < 1$. This is also the condition for the convergence of chaotic iterations of Chazan and Miranker [CM69].

In order to analyze the total cost of asynchronous methods one must study the speed of convergence and consequently determine the number of steps necessary to reduce the truncation error of the initial approximation until the needed threshold is reached. It is possible to prove that the number of steps needed to reduce the truncation error to the value 2^{-t} has a lower bound given by the following expression:

$$-\left[\lim_{j\to\infty} \inf \left(\frac{j}{p_j}\right)\right] \frac{t}{\log \rho(A)}$$

where p_j is the maximum number of times that the components have been updated for $k \leq j$. The inverse of the expression in square brackets (which is 1 in classical iterative methods) provides a measure of the percentage of "active" approximations.

Bibliographic notes

Due to the quantity and quality of the results obtained, parallel algorithms are now one of the most important fields of computer science. Many important parallel algorithms have been developed, which are substantially new (i.e. they are not simply parallel versions of known sequential algorithms). Some of them have a purely theoretical interest, while others can be actually implemented on real, or realistic, parallel machines.

Several important results, concerning the concrete implementation of efficient algorithms, have been published in the journal *Parallel Computing*.

Regarding the more theoretical aspects, among the parallel algorithms to solve arithmetical problems, we point out the work by P.W. Beame, S.A. Cook and H.J. Hoover, "Log Depth Circuits for Division and Related Problems", [BCH84], where the proof that division is in NC^1 is given. In the field of combinatorial algorithms, two recent reviews are of particular importance [KR90], [EG88]. In the field of parallel algorithms for numerical linear algebra, the work of Csanky, which is published in "Fast Parallel Matrix Inversion Algorithms" [Cs76], is fundamental. The importance is due to its implications to several other computational problems. Other significant contributions can be found in [He78], [Be84], [BGH82], [Mu86], [PR85].

Problems

4.1 Apply *divide et impera* to invert a triangular matrix. Which bounds on T and H are obtained (see Section 4.1.1)?

4.2 Apply the method of vector iterations to general matrices. What is the drawback of this approach?

4.3 Given a parallel algorithm described in terms of arithmetic circuits, find out when the technique of reduction of the used hardware cannot be applied.

4.4 Use the arithmetic algorithms of Section 4.2 to obtain algorithms for operations on polynomials (sum, product, division). Which performance is obtained?

4.5 Implement the mergesort algorithm for sorting on a PRAM.

4.6 Give a PRAM version of Gaussian elimination with partial pivot. Is it possible to obtain $T(n) = O(n\log n)$?

4.7 Implement the Csanky algorithm on a PRAM and say if the bound $O(\log^2 n)$ can be maintained.

4.8 Say if the iterative methods for solving linear systems have efficient EREW PRAM implementations. State the reasons for your answer.

Chapter 5
Parallel Architectures

In the previous two chapters we introduced and developed concepts and methods of a mostly theoretical nature. The main purpose was to introduce the reader to the fundamentals of parallel computing: computational models, with their relative measures of complexity (Chapter 3), and parallel algorithm analysis with respect to the use of critical resources, as detailed in the models (Chapter 4).

In this chapter we will describe a different approach to the study of algorithms, and to parallel computing in general, as opposed to the view presented in Chapter 3. This approach is based on actual models, instead of on abstract ones, consisting of real machines or of machines that can be possibly built. For this reason we will illustrate the characteristics of a few parallel architectures, and we will analyze their working principles. In particular, we will study implications on the structure of computations and on classes of problems which can be efficiently treated. On the other hand, we will not go to great depths in topics involving technology, nor will we deal with "systems" aspects (scheduling, memory management, etc), for which we suggest specialized textbooks such as [HWB84].

A study of complexity based on actual models enables one, among other things, to evaluate critically advantages and disadvantages of the classical complexity theory of parallel computing which is founded on abstract models. According to some authors, the "concrete" approach we will discuss in this chapter is the only one with a sensible meaning, since many of the assumptions we make on abstract models cannot be satisfied in any way, and can lead to questionable results. One of the criticisms concerns the fact that it is not realistic to consider, in determining the result of a computation, only logical and/or arithmetic operations while neglecting data access and communication [GEN78], [FIS88]. Note that data access and communication problems are also present (even if they have a lesser effect on performance) in the sequential case. However, the classical complexity theory, which ignores them, has given results of great interest, the utility of which is beyond discussion. The abstract and concrete approaches are simply complementary. In the first case the primary objective is the analysis and the synthesis of efficient algorithms with respect to actual models. In the second case the objective is seeking the least possible parallel time (that is, the maximum level of parallelism). This search forces one to make concessions to the computational model. Even if the first approach is clearly better suited to present technology, it is reasonable to think that in the future low asymptotic cost algorithms will play an important role [BGH82].

The "algorithmic" emphasis of the present study will in any case considerably restrict the class of parallel architectures we will be concerned with. The fundamental prerequisite is that communication between processors participating in the computation take place in "reasonable" times (but, as outlined before, not in negligible times). This has immediate consequences both on processor distance and on the characteristics of the communication system (presence of a global memory, topology and transmission bandwidth of the communication system). In our investigation we will then exclude parallel architectures like local and geographic networks, and all those parallel architectures the primary purpose of which is resource sharing and/or global variable optimization (processor use, user waiting times, etc) rather than the efficient execution of parallel algorithms.

The present chapter is structured in the following way. Section 5.1 will describe the main architectural characteristics of parallel machines. It will also discuss a parallel machine classification scheme centred on a programmer's point of view [FLY66, FLY72]. Finally, it will analyze the main features of synchronous and asynchronous parallel programming. In Section 5.2 we will focus our attention on interconnection structures which can be used to link processors among themselves in distributed computing environments. Finally, Sections 5.3 and 5.4 will examine a few characteristics of architectures with and without global memory respectively.

5.1 Introduction to parallel architectures

We are interested in a description of real parallel machines providing, in analogy with what we have seen in the case of abstract models, direct indications on the structure of computations. However, in contrast to the former case, this description must also consider data access and communication aspects. The discussion will be conducted at the *firmware architecture* level. In a computational system we have several important description levels, each of which requires a particular structural and functional view. Also, each level is characterized by a special language used to program its behaviour.

At the firmware level we have two types of processing unit: processor and memory. These units are linked, and interact during a computation by means of a visible structure. The firmware level behaviour of a computer can be programmed using machine language. For clarity though, we will use, while describing computations, higher-level languages, for instance a symbolic machine language or the usual Pidgin ALGOL. All the mechanisms to be used in the programs (for example processor-to-processor communication functions) must be thought of as *primitive* at the firmware level, and therefore they must be implemented at a lower level, by means of microprograms for instance.

At the firmware architecture level the input/output system is also visible. We will not take it into account however since we assume data and program to be already present in memory at the start of the computation. We also assume that there is no data communication with the outside world during computations.

Concerning processor and memory as processing units, we have the following general considerations. A processor is any unit capable of executing arithmetic/logic operations, and, in addition, data transfer and routing. This definition might not coincide with the normal idea of a processor. For instance, in general it is not required of the processor to be able to perform functions such as instruction call and decoding, or to be able to generate its own instruction flow.[1]

The common idea one has of memory is that of an information "container", that is, of something that is simply used by entities able to process information. In fact, we say that a processor writes to, or reads from, memory a datum. This view, useful at some level, for instance while describing sequential computations, is not much of a help in understanding some problems related to a parallel computing environment. Actually, memory is a unit also capable of performing some functions. In particular it can communicate with other units, and it can transfer, when requested, the contents of a location it controls. Therefore, when speaking of a processor accessing a memory for reading we imply a sequence of the following type:

---

[1] Note that each processor in the abstract PRAM and network models seen in Chapter 3, although with a limited repertory, could generate its own instruction flow, independent from the other processors.

- the processor sends, by means of some communication system, a message specifying the requested operation (read) and the address of the location it wants to access;
- the memory unit, once it has received the message, decodes it and executes the request;
- the memory unit sends the contents of the location to the requesting processor.

Whenever possible however we will treat the memory-processor interaction as an information access problem. Finally, note that a memory cannot execute arithmetic/logic operations of a general type, nor can it perform data routing.

5.1.1 Pipeline parallelism

Parallelism can be introduced within a single processor by means of a technique known in industrial applications even before the first computer had been constructed: the assembly line. In an assembly line several pieces are worked on simultaneously, while, at a given instant, each piece is given a different treatment. The first *supercomputers* for scientific computations (the CRAY-1 and the CDC Cyber 205 for instance) reflected this kind of organization. The structure of the computing unit of these machines, known as *pipeline,* has been until now the most efficient for scientific computations. To better understand this form of parallelism it is useful to refer to a few notions on number representation and machine arithmetics (see Appendix B).

The pipeline organization of the working units of vectorial computers such as the CRAY-1 allows one to exploit the decomposition of arithmetic instructions in different phases (see Appendix B). The execution of a vectorial instruction such as:

$$C[1 \ldots N] \leftarrow A[1 \ldots N] \text{ op } B[1 \ldots N]$$

where $\text{op} \in \{+, -, *, /\}$, can be indicated as in Figure 5.1. At instant 1, the unit applies phase 1 (computation of the difference between exponents) to the elements $A[1]$ and $B[1]$; at instant 2 it aligns the mantissas of $A[1]$ and $B[1]$ and computes the difference between the exponents of $A[2]$ and $B[2]$, and so on. It is clear that in the steady state, that is, since the moment the new value of $C[1]$ has been computed (instant 4 in Figure 5.1), and until all data have been processed, at each instant the working unit gives a new result. Typically, an "instant" in time corresponds to a basic machine loop. Therefore, after an initial delay, the working unit gives a result at each loop,

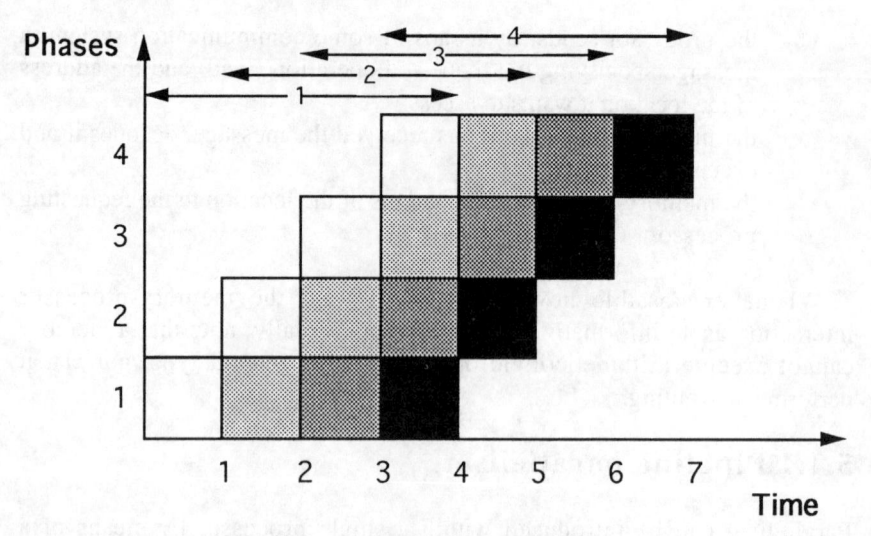

Figure 5.1 Pipeline processing.

even if each pair of data is present for a period of four loops. This fact translates into a speed-up which asymptotically converges to the value 4.

The parallelism in these machines manifests itself by a superposition of different phases of the same instruction on different data.

In the presence of a parallelism of this kind we speak of *vector processors* and therefore of vector machines, since the optimal situation is that when the pipeline arithmetic is fed with vector data.

The most recent computers for scientific computation, such as the CRAY X-MP or the IBM 3090-VF, can consist of one or more processors, each of which is built using the pipeline technique, accessing a shared memory. On a more macroscopic level they are then multiprocessors with a common memory. Since these processors are very sophisticated and expensive, their numbers in a multiprocessor are severely limited (at most 4 in the CRAY X-MP).

5.1.2 Classification according to Flynn

From a programming point of view, one of the most important classifications of parallel machines has been proposed by Flynn [FLY66, FLY72]. The discriminating concepts are *instruction stream* and *data stream*. By instruction stream we mean a sequence of instructions executed by a processing unit (processor). By data flow we mean the sequence of operands it manipulates. Controlling instruction and data flows means determining the order in which the program instructions will be executed by a processor, and on which data they will operate. Such control is made possible by the programming language structures. In the case of a sequential machine there are a single instruction flow and a single data flow. According to the Flynn classification a sequential machine is then called a *SISD* (*Single Instruction stream Single Data stream*) machine.

Singling out the key concepts of instruction and data flow leads us logically to determine three classes of parallel machines, independent of the fact that these machines can actually be built or have any practical interest:

1. Machines where a single instruction flow is applied to several data flows. Such machines are called SIMD (*Single Instruction stream Multiple Data stream*) type parallel machines.
2. Machines of the MISD (*Multiple Instruction stream Single Data stream*) type, where multiple instruction flows are applied to the same data flow.
3. Machines of the MIMD (*Multiple Instruction stream Multiple Data stream*) type, where multiple instruction flows are applied to several data flows.

The SIMD and MIMD classes proved themselves not only as a way of classifying the existing architectures, but also a widely used conceptual tool to analyze or propose new parallel architectures. Further, the near totality of actually built parallel machines falls within the above classes.[2] The same is not true for the MISD class. In fact, although it is possible to envisage situations where several instructions must simultaneously operate on each single datum (for an example see [STO75], Chapter 8), this is in general an unrealistic model for parallel computation.

5.1.3 SIMD parallelism

Programming SIMD machines is also called *synchronous parallel programming*. There is in fact a single program, and, differing from the case of the PRAM and network models, all processors are constrained to execute the same instruction.[3] In this way the parallel flows, automatically generated

---

[2]The classification given by Flynn is clearly exhaustive with respect to the key concepts of instruction and data flow. There are however some approaches to parallel computation where these concepts do not find immediate application. We are referring in particular to computational models completely different from the ones discussed in the present book, such as the data-flow model. As a consequence we prefer excluding from the Flynn classification architectures deriving from these different models.

[3]In fact, since every processor in a PRAM or in a network is able to generate its own instruction flow (each processor has its own program counter register) the model resembles more a MIMD machine rather than a SIMD one. As in the case of MIMD machines in a

172 Parallel architectures

by an instruction on different data, rejoin each other when the instruction itself has been completed. This is done by mechanisms outside the control of the programmer.

Vector processors are a type of SIMD machine. In these processors a single vector instruction controls the simultaneous (although in different phases) manipulation of several data. Despite the fact that vector processors represent the largest computing power now available to the scientific user, this power stems from a very sophisticated (and very expensive) technology, rather than from an extensive use of parallelism. Such a fact makes vector computation less interesting from the point of view of algorithm analysis and synthesis.

The following two examples illustrate one of the key problems found in the implementation of efficient programs for a SIMD computer.

Example 5.1 (data dependence [PAW86])
Consider the following code fragments:

P_1	P_2	P_3
S_1: $A \leftarrow B+C$;	S_1: $A \leftarrow B+C$;	S_1: $A \leftarrow B+C$;
S_2: $D \leftarrow A+2$;	S_2: $B \leftarrow D/2$;	S_2: $D \leftarrow A+2$;
S_3: $E \leftarrow A*3$;		S_3: $A \leftarrow E+F$;

In fragment P_1 instruction S_2 and S_3 use the value of A determined by instruction S_1. For this reason S_2 and S_3 cannot be executed before S_1. The dependence of S_2 (and of S_3) on S_1 is called *flow dependence*, since data flow goes from S_1 to S_2. This dependence is denoted by $S_1 \delta S_2$.

In fragment P_2 instruction S_1 uses an old value for B and therefore it must be executed before S_2, which on the other hand modifies such a value. This relation is called *antidependence* and is denoted by $S_1 \delta^- S_2$.

In fragment P_3 instructions S_1 and S_3 modify the value of A. If S_3 had been executed before S_1, instruction S_2 would use an incorrect value of A.. This dependence is called *output dependence* and is denoted by $S_1 \delta^\circ S_3$.

Example 5.2 (loop vectorization [PAW86])
The dependence relations existing between the instructions in a program can be immediately used to generate efficient code for vector computers. In particular, they are useful in establishing whether, among the instructions

---

PRAM the exercise of some care is required to synchronize the activity of different processors (see Example 3.31). In a MIMD machine however the additional constraint of the single program is released.

making up an iterative instruction, there are dependences preventing the transformation of a loop into a single vector instruction.

Consider the following program:

for $i \leftarrow 2$ **until** N **do begin**
    S_1: $C[i] \leftarrow A[i]+B[i-1]$;
    S_2: $E[i] \leftarrow C[i+1]$;
    S_3: $B[i] \leftarrow C[i]+2$
end

To determine the dependence relation existing among the various instructions it is advisable (unless one has great experience in this kind of problem) to "unfold" the iterative instruction by writing the corresponding program in a straight line. In our case we have:

S_1: $C[2] \leftarrow A[2]+B[1]$;
S_2: $E[2] \leftarrow C[3]$;
S_3: $B[2] \leftarrow C[2]+2$;
S_1: $C[3] \leftarrow A[3]+B[2]$;
S_2: $E[3] \leftarrow C[4]$;
S_3: $B[3] \leftarrow C[3]+2$;
...

Thus, we find that the following relations hold: $S_2 \; \delta^- \; S_1$ (since S_2 uses the old values contained in vector C while S_1 modifies them), $S_1 \; \delta \; S_3$ (trivial) and $S_3 \; \delta \; S_1$ (since S_1 uses the values determined by S_3). One immediately notes the presence of the cyclic dependence $S_1 \; \delta \; S_3$, and $S_3 \; \delta \; S_1$ (in the computation graph there would be a loop). This means that the iterative instruction cannot be transformed into three vector instructions:

S_1: $C[2 \ldots N] \leftarrow A[2 \ldots N]+B[1 \ldots N-1]$;
S_2: $E[2 \ldots N] \leftarrow C[3 \ldots N+1]$;
S_3: $B[2 \ldots N] \leftarrow C[2 \ldots N]+2$;

since these give different results.

Consider now the following program:

for $i \leftarrow 1$ **until** N **do begin**
$\quad S_1: A[i] \leftarrow B[i];$
$\quad S_2: C[i] \leftarrow A[i]+B[i];$
$\quad S_3: E[i] \leftarrow C[i+1]$
end

It contains the data dependences $S_1 \,\delta\, S_2$ and $S_3 \,\delta^-\, S_2$. Since there are no cyclic dependences, the iterative instruction can be completely vectorized, though with the care of executing S_3 before S_2 (because $S_3 \,\delta^-\, S_2$). In this way the sequential program is equivalent to the following vector program:

$S_1: A[1\ ...N] \leftarrow B[1\ ...N];$
$S_3: E[1\ ...N] \leftarrow C[2\ ...N+1];$
$S_2: C[1\ ...N] \leftarrow A[1\ ...N]+B[1\ ...N];$

There are at least two ways to implement high-level languages for vector computation. The first is to extend a sequential language by means of vector instructions of the kind used until now. In this case the programmer must make sure that the vector code has a correct behaviour (that is, it is equivalent to the sequential version of the algorithm possibly developed before). The second solution prescribes that the language compiler must find dependence relations and translate iterative instructions into vector instructions (this time in machine language). In this case the programmer still writes sequential code. Both solutions have advantages and disadvantages. In the first instance the programmer bears, theoretically, a greater workload and is liable to make mistakes in the vectorization, while having the possibility of writing a very efficient code. In the second case the main advantage is the possibility of using on vector machines existing code written for sequential machines. The disadvantage is that the compiler, since it must guarantee the absolute equivalence of the programs, usually cannot vectorize the code completely.

The solution employed by almost all supercomputers is the vectorizing compiler. They have in fact at least one FORTRAN compiler (the most widely used language in scientific computations). Furthermore, as an alternative to the language extension, the working environment usually has available a utility subprogram library (for instance the ESSL library present on the IBM 3090-VF) made of highly optimized code.

In vector processors the SIMD parallelism manifests itself by means of the *phase superposition* of an instruction on different data. The other approach, more immediate perhaps, consists of the *concurrent application* of an instruction to distinct data. This type of parallelism can be obtained by means of a SIMD architecture, called *array processor*.

Figure 5.2 schematically shows the logical organization of an array processor. The control unit generates the instruction flow, that is, it executes the instruction recall from memory and decoding functions. It further executes some of the instructions, jump instructions in particular, which are an intrinsic part of the control. The N working units can all be identical, or they can be specialized to execute particular instructions. The input (output) data flow into (out of) the working units can originate from (be directed to) the local memory or from (to) another working unit. Note then that an array processor is a distributed memory multiprocessor.

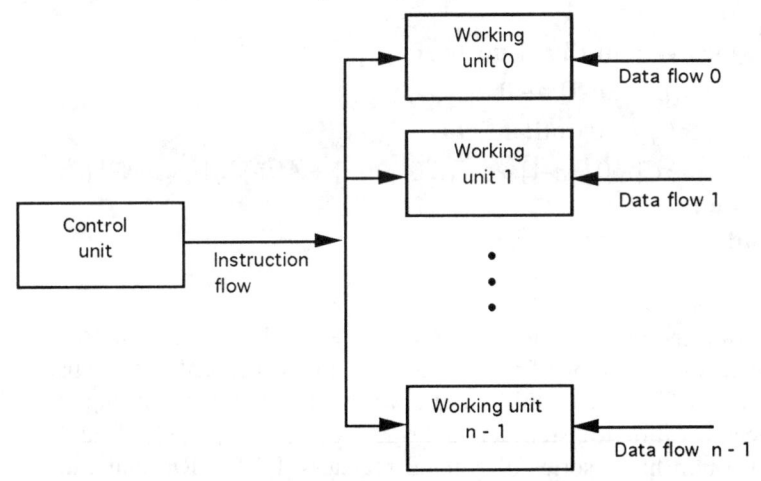

Figure 5.2 Array processor structure.

In the simplest case one can think that the working units, all the same, are called only in the case of vector instructions. For instance, the vector instruction:

$$C[0 \ldots n-1] \leftarrow A[0 \ldots n-1] + B[0 \ldots n-1]$$

is executed in the following way. The instruction is decoded by the control unit. This unit subsequently distributes to each processor the indication on the type of instruction and information to generate the appropriate data flow. In particular the working unit i, $i = 0, \ldots, n-1$, generates the input flow $A[i]$, $B[i]$, $A[i+n]$, $B[i+n]$, ..., $A[i+kn]$, $B[i+kn]$, and the output flow $C[i]$, $C[i+n]$, ..., $C[i+kn]$, $k = \lfloor (n-i)/n \rfloor$.

176   Parallel architectures

Example 5.3 (multiplication of order n matrices)
Many sequential programs containing iterative loops (**for** construct) can be transformed into efficient programs for SIMD machines. The case of matrix multiplication is a good example. Compared to the classic sequential scheme, made of three nested loops (giving a cost $O(n^3)$), in the following SIMD program the innermost loop is transformed into a vector instruction. Assuming, for the sake of simplicity, that the number of working units is not less than the order of the matrices, the time cost of the SIMD algorithm is $O(n^2)$.

```
begin
    for i ← 0 until n-1 do begin
        C[i, 0 ...n-1] ← 0;
        for j ← 0 until n-1 do
            C[i, 0 ...n-1] ← C[i, 0 ...n-1] + A[i,j]*B[j, 0 ...n-1]
    end
end
```

To understand how an array processor works it is interesting to see a few examples of assembler instructions that could reasonably be a part of its repertory [STO75]. To this end we will assume that the working unit k has an accumulation register, denoted by ACC[k], $k = 0, ..., N-1$, and that the control unit has a series of general registers, R1, ..., Rn, that can also be used as index registers. The matrices A and B to be multiplied, also like the product matrix C, are stored by columns, that is, the ith columns of A, B and C, are stored in the local memory of the working unit i, $i = 0, ..., N-1$.

Table 5.1 describes some instructions operating on vectors. Note that the construct $A[Ri]$, decoded by the control unit and passed to the working units, indicates the element of index h of the kth column of A, where h is the value contained in the register Ri, and $k = 0, ..., N-1$. In other words, each working unit k interprets the construct $A[Ri]$ as $A[Ri,k]$.

The BCAST instruction allows a working unit to communicate the value of the contents of its accumulator to all other units. Its presence in the instruction repertory requires, for efficiency reasons, the existence of a distributing connection among the various units. This connection can be implemented as a low-cost structure (a bus for example).

Introduction to parallel architectures   177

Instruction	Operands	Description
LOADV	constant	$acc[k] \leftarrow$ constant, $k = 0, ..., N-1$
LOADV	A[Ri]	$acc[k] \leftarrow A[Ri,k]$, $k = 0, ..., N-1$
STOREV	A[Ri]	$A[Ri,k] \leftarrow acc[k]$, $k = 0, ..., N-1$
ADDV	A[Ri]	$acc[k] \leftarrow acc[k] + A[Ri,k]$, $k = 0, ..., N-1$
MULV	A[Ri]	$acc[k] \leftarrow acc[k] * A[Ri,k]$, $k = 0, ..., N-1$
BCAST	Ri	$acc[k] \leftarrow acc[Ri]$, $k = 0, ..., N-1$

Table 5.1 Assembler SIMD instructions.

Example 5.4

Using the instructions in Table 5.1 one can write the following assembler version of the program to multiply two order N matrices (where it is assumed that the value N is already stored in register R0 of the control unit). Note that the program also uses a few simple scalar instructions (a conditional jump instruction and instructions to load and increment index registers, to be exact). As already remarked, these instructions are directly executed by the control unit.

	Instruction		Comment
	LOAD	R1,0	/* R1 contains index i
	LOAD	R2,0	/* R2 contains index j
LOOPI	LOADV	0	/* $acc[k] \leftarrow 0$
	STOREV	C[R1]	/* $C[i,k] \leftarrow 0$, $k = 0, ..., N-1$
LOOPJ	LOADV	A[R1]	/* $acc[k] \leftarrow A[i,k]$, $k = 0, ..., N-1$
	BCAST	R2	/* $acc[k] \leftarrow acc[j] = A[i,j]$, $k = 0, ..., N-1$
	MULV	B[R2]	/* $acc[k] \leftarrow A[i,j] * B[j,k]$, $k = 0, ..., N-1$
	ADDV	C[R1]	/* $acc[k] \leftarrow C[i,k] + acc[k]$, $k = 0, ..., N-1$
	STOREV	C[R1]	/* $C[i,k] \leftarrow acc[k]$, $k = 0, ..., N-1$
	INC	R2	/* $j \leftarrow j+1$
	JLE	R2,R0,LOOPJ	/* if $j < N$ goto LOOPJ
	INC	R1	/* $i \leftarrow i+1$
	JLE	R1,R0,LOOPI	/* if $i < N$ goto LOOPI
	...		

We have already seen that the input data flows into a working unit can originate from the local memory or from another working unit. For instance, the instruction LOADV activates N data flows from memory to working units, while the instruction BCAST activates $N - 1$ data flows from one unit to the remaining $N - 1$. In the schematics shown in Figure 5.2, memory and

178 Parallel architectures

connections are not shown. Figure 5.3 shows a more complete situation. The interconnection structure is almost always a network, called *permutation network*, since it allows, after the execution of a communication instruction, a data permutation among processors. A somewhat usual situation for this kind of machine is that the working units are connected according to a *mesh* topology.

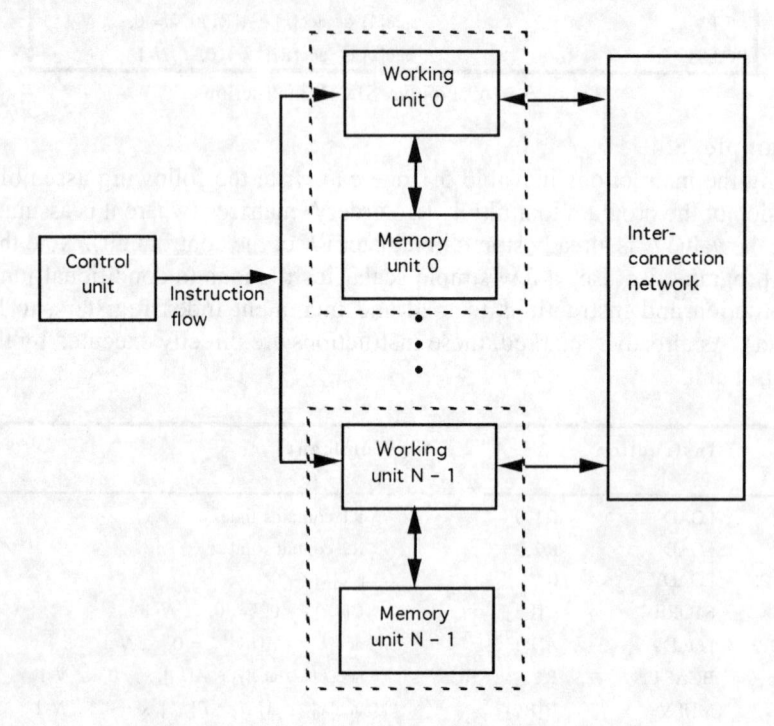

Figure 5.3 Distributed memory SIMD multiprocessor.

Note that the high regularity of array processors is very well suited to their implementation with VLSI technology (see Section 6.1), when single processors are sufficiently simple. In this way a high degree of parallelism can be obtained. A few examples of machines falling into this class are the Illiac IV (perhaps the first true parallel machine ever built), which had 64 working units capable of executing arithmetic instructions of a complete repertory, the Distributed Array Processor, and the Massively Parallel Processor. These last two had 4096 and 16384 processors respectively, but with computational capabilities restricted to the simplest operations. All the above machines had a distributed memory.

As we saw in the example regarding matrix multiplication, array processors are well suited, like vector processors, to the solution of problems where the main computational part consists of the processing of a highly regular data structure, typically an array with one or more dimensions. There are several problems which give rise to computations of this kind, especially in the field of numeric linear algebra.

A fundamental difference between vector processors and array processors is the presence in the latter of a data communication aspect. In the computation done by an array processor the results given by a processor can be moved, under program control, to another unit for further processing. In this case, the interconnecting structure topology (that is, the allowed special data permutations) is of fundamental importance in the implementation of efficient computations (see Section 5.2).

5.1.4 MIMD parallelism

In the case of MIMD parallelism (*Multiple Instruction stream Multiple Data stream*) the programmer can control several instruction flows and several data flows. An instruction flow executed by a real processor is called *virtual processor, process* or *task*. Note that all processes can consist of instructions which belong to physically different programs. This is the reason why the abstract multiprocessor models seen in Chapter 3 (where each processor executes instructions from the same program) cannot be classified as MIMD parallelism models. This high flexibility is an indication that MIMD parallelism could be more general than the SIMD one. In fact, a problem without structure regularity, but having potential parallelism, can be subjected to a MIMD type processing. The greater flexibility carries a price tag, since some problems arise which are absent in the SIMD case. In particular these are synchronization problems, at the process and data levels, and processor allocation problems.

In the SIMD case, (real) processor synchronization is guaranteed by the presence of a single instruction flow. Thus, all processors simultaneously start and end the execution of any instruction. Processor synchronization using instruction flow also eliminates the data level synchronization problem. The synchronization is indeed guaranteed by the implementation of the communication instruction itself.

In the MIMD parallelism case, synchronization must be ensured in another way. Processes, being independent instruction flows (possibly generated by different programs), constitute mutually asynchronous activities. It is apparent that, as in the SIMD case, special instructions must exist to ensure a correct data transfer among communicating processes. However, we cannot assume that these instructions are simultaneously executed by the communication partners. It is this fact that makes process handling by the system, and process programming by the user, more complex.

There is also another aspect which gives higher flexibility to the MIMD model, but that increases programming difficulty. It concerns the possibility of dynamically creating processes. The processes, being "virtual processors", are not constrained to be less than or equal to the real processors in number. The hidden danger, in this case, is in the need to control the theoretically unlimited number of distinct instruction flows which can be generated during a computation.

Direct process interaction

The first aspect we take into account is that of *direct interaction* among processes. The term direct refers to the fact that a process is aware of the existence of the processes with which it interacts. There is also a form of *indirect interaction* characterizing situations where several processors (real or virtual ones) get into a conflict while accessing a common resource (a memory module for instance). We will not concern ourselves here with indirect process interaction.

One has direct interaction when two processes want to communicate. The situation can be modelled in two different ways, according to the logic architecture of the machine, whether it is with common or distributed memory. Let us first consider the case of a common memory architecture. The communication scheme is very simple. The sender process writes a message in a preset memory area and subsequently the receiver(s) reads it. Let us see the problems arising in this respect.

Writing data to memory requires, in general, several communications between the processor requesting communication (or the process executing on it) and one or more memory modules. If this is the case, there is the possibility that a second (receiver) process can enter these communications, acquiring as a consequence incomplete or meaningless information. This occurrence is due to the fact that processes are asynchronous activities. In the common memory MIMD model there must be instructions to synchronize two processes:

(a) from a physical point of view, with respect to access to the same variable;
(b) from a logical point of view, with respect to the meaning of the contents of such a variable.

To this end, one associates with a shared variable a *condition flag,* or *event,* and one defines two instructions, WAIT and SIGNAL, which operate on condition flags. The function of the instruction WAIT(E), executed by some process, is to check whether some event E has taken place. If yes, the process can continue, otherwise its execution must be somehow paused. The instruction SIGNAL(E) mirrors the previous instruction. It notifies that event E has taken place, and, further, that one of the processes which were (possibly) paused waiting for such an event can be reactivated. Note that which event is represented by which flag is decided by the programmer through the use of the flag itself. From an implementation point of view, a condition flag can simply be an integer variable. If this is the case we assume that $E \geq 1$ indicates the number of events, of the type associated with this flag, which have taken place. $E \leq 0$ on the other hand will indicate that the event has not taken place, and, further, that precisely $-E$ processes have been paused waiting for it. The above arguments lead to the following definitions of WAIT and SIGNAL.

procedure WAIT(E);
begin
    $E \leftarrow E-1$;
    if $E < 0$
        then "pause process"
end

procedure SIGNAL(E);
begin
    $E \leftarrow E+1$;
    if $E \leq 0$
        then "activate one of the paused processes"
end

Note that an operation such as "$E \leftarrow E-1$" requires two accesses by the process to the location denoted by E: reading the value in (that is, loading it into one of the registers of the processor executing the process) and subsequent writing back the modified value. It is essential that no other process enters to read and/or modify the value of E. Otherwise, cases such as the following could arise.

1. Process A starts execution of WAIT, that is, it reads the value of E, let it be 1, and stores it in one of its registers, let it be R0.
2. Process A decreases the value of R0, which becomes 0.
3. Process B starts execution of SIGNAL, that is, it reads the value of E, which is still 1, and stores it in its own R0 register.
4. Process A writes the value in R0 to E, so now E contains the value 0.
5. Process B increases the value of R0, which becomes 2.

6. Process *B* writes the value of R0 to *E*, so now *E* contains the value 2.

The result is that the final value of *E* is 2, while it should be 1 (since WAIT decreases and SIGNAL increases). It is apparent that the contribution of WAIT has been lost. If the two operations were indivisible, the execution sequence would have the following enforced order: 1, 2, 4, 3, 5, 6, which gives the correct result. The indivisibility of some special instructions is ensured at the level of function definition of the processor and of the memory units themselves.

By having instructions such as WAIT and SIGNAL, it is possible to program the communication between processes while satisfying requirements (a) and (b) above. Consider the case of symmetric communication (from sender to receiver). Assume that the message is written to a variable *V*, which has associated with it the two condition flags, PRESENT and ABSENT, initialized with the values 0 and 1 respectively. The following program fragments show a correct use of the WAIT and SIGNAL instructions.

process *A*	process *B*
... ...	
"produces a message for B"	...
WAIT(ABSENT);	WAIT(PRESENT);
"writes the message to *V*"	"read the message in *V*"
SIGNAL(PRESENT);	SIGNAL(ABSENT);
...	"use the message received from *A*"
... ...	

Process *A*, before writing the message to *V*, executes WAIT on ABSENT. If ABSENT = 1, then *V* is "logically" empty, that is, its contents are meaningless, and *A* can proceed by writing the message and signalling its presence with a SIGNAL on PRESENT. Note that while *A* writes, *B* cannot "intrude" and read the contents of *V*, since PRESENT = 0. The case ABSENT = 0 means that a previous message written to *V* has not been read yet, and therefore process *A* must wait until *V* has been "emptied". For its part, process *B* behaves as a mirror image. If PRESENT = 1, then the message is present in *V*, and *B* can read it, signalling afterwards that the contents of *V* are now meaningless. Also in this case, while *B* reads ABSENT = 0, and therefore *A* cannot intrude to write another message. If, on the other hand, PRESENT = 0, then *B* must wait for the message to be written.

Let us now see a complete example.

Example 5.5 (sum of $N = 2^n$ numbers)
Let us assume we want to add the elements of a vector A of order N on a common memory MIMD machine. To this end we use the classic fan-in associative algorithm which, in the presence of at least $N/2$ processors, has a cost of $O(\log N)$ steps. N processes take part in the computation, every one of which has stored in the variable k an identifier number between 0 and $N - 1$. We associate with each variable $A[i]$ a condition flag $E[i]$, $i = 0, ..., N - 1$, with the understanding that, if $E[i] = 1$, the value of $A[i]$ can also be used by a process different from process i.

```
...
if (k mod 2) = 0 then E[k] ← 0
else E[k] ← 1;                  comment flag initialization;
for i ← 2 step i until N do
   if (k mod i) = 0 then begin
      WAIT(E[k+i/2]);
      A[k] ← A[k]+A[k+i/2];
      SIGNAL(E[k])
   end
```

Note the use of a variable step iterative instruction (**for** $i \leftarrow$ 2 **step** i **until** N **do**), executed $n = \log N$ times. The core of this instruction is executed the first time by all processes for which the value of k is even, the second time by all processes for which k can be divided by 4, and so on. The last step is executed only by the process for which $k = 0$.

The presence of instructions such as WAIT and SIGNAL allows the solution of the communication problem among processes, ensuring in the meantime data synchronization (that is, exclusive access to shared variables). It also allows the programmer to synchronize processes (reading only meaningful messages, waiting in the case of message absence). This last aspect however is not automatic. In other words, there are WAIT and SIGNAL instructions, but they must be correctly used. It is clear that a process executing a WAIT for which the corresponding SIGNAL is never executed remains in a wait state indefinitely. In the case of "complicated" programs it is not obvious that everything will work correctly.

There are also more delicate problems. Assume that process A produces a datum to be sent to B and that, at the same time, it is waiting on the latter for some result. If both processes are supposed to wait for data from the other process before sending their own, the combined wait state cannot be unlocked. In this case the order of execution (rather than a missing execution) is the source of problems. Note that *cyclic dependences* of this kind can involve more than two processes (A waits for data from B, which waits for data from C, ..., which waits for data from A) and can lead to a situation known as *deadlock*.

Let us now give a precise meaning to the sentence "if condition E is not verified process A goes into a wait state". What is required is that all information making up the context of process A (contents of general registers, contents of the program counter, etc) be stored in a data structure, called a *waiting queue,* associated with the flag E. When another process executes SIGNAL on E, if the context of A is the first in the waiting queue, A can continue execution. Obviously, a real processor must be available to load the context of A. The great advantage of a solution of this kind is that the number of processes is decoupled from the actual availability of real processors. A given algorithm can be logically subdivided into several parallel activities, and the effect of having more or less real processors (although important) concerns only its completion time. This schematization also entails the division of processes into *executing* processes, *waiting* processes and *ready for execution* (or simply *ready*) processes. Executing processes are those currently having assigned to them a real processor for execution. Waiting processes are those having their context stored in the queue associated with a condition flag. Finally, ready processes are those which could theoretically continue execution, but are waiting for the most important resource, that is, a real processor. The diagram below shows the possible states of a process and the events causing transitions from one state to the other.

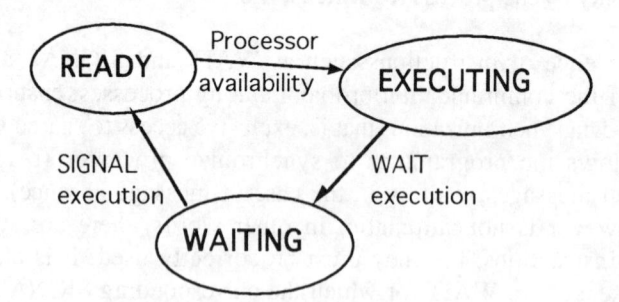

An analysis in some ways similar to the one we have just completed can be done for distributed memory multiprocessors. In this case there are no access problems for shared variables, but there is the problem of a correct synchronization among the interacting processes.

In this model the fundamental communication instructions are SEND and RECEIVE. A process A needing to communicate a datum to process B will execute a SEND instruction, giving as parameters at least the destination process and the location where the datum to be transmitted is contained. Process B will execute a RECEIVE instruction, indicating the local variable where the message will be stored and, possibly, the sender process.

process A	process B
...	...
$V \leftarrow$ "message"	...
SEND(B,V);	RECEIVE(W, A);
...	"use the contents of W"
...	...

Since activities are *asynchronous*, when process A executes the SEND primitive, process B can be at any point within its program. It is apparent that the implementation of the SEND and RECEIVE instructions must take this fact into account. We have two possible cases:

1. B executes RECEIVE before A executes SEND. In this case process B execution must be paused until data are available. Similarly to what happened with the WAIT instruction, the process must go into a wait state.
2. B executes RECEIVE after A executes SEND. In this case the message must be stored "somewhere" in order to be able to transfer it later to the location specified by B in the RECEIVE instruction.

The behaviour of the SEND and RECEIVE instructions ensures that the data transfer takes place correctly. We are left, as in the case of WAIT and SIGNAL, with the problem of a correct use of the instructions. In particular, we must avoid the occurrence of deadlock situations.

Example 5.6 (sum of $N = 2^n$ numbers)
Let us consider once again Example 5.5 and solve the same problem on a distributed memory MIMD machine. In this case also N processes, each one having stored in the local variable k an identifier number between 0 and $N-1$, take part in the computation. Communications use the identifier P_k to indicate the kth process ($k = 0, ..., N-1$). Data distribution is such that in the local memory of process P_k there is contained the element $A[k]$ of the vector, the elements of which must be added. TEMP is a temporary variable present in the local memory of every process.

...
for $i \leftarrow 2$ **step** i **until** N **do**
   if $(k \bmod i) = 0$ **then begin**
      RECEIVE(TEMP,$P_{k+i/2}$)
      $A[k] \leftarrow A[k]$+TEMP **end**
   else if $(k \bmod i) = i/2$ **then** SEND($P_{k-i/2}$,$A[k]$)

Let $N = 8$. In the first step, the processes for which $k = 1, 3, 5$ or 7 send the value $A[k]$ to processes P_0, P_2, P_4 and P_6 respectively. In the second step, the processes for which $k = 2$ or 6 send the value $A[k]$ to processes P_0 and P_4 respectively. In the last step, process P_4 sends the value $A[4]$ to process P_0 for the final sum.

Dynamic process creation

The other great problem facing the MIMD programmer is process creation and termination. We have already seen that every process can execute instructions belonging to different programs. This gives a great flexibility, but, in the terminology of Chapter 3, it makes the model "little uniform", that is, the model cannot be easily described. Imagine, for instance, the need to write programs for 10 000 different processes!

Generally speaking, the programmer writes a single program text, in which a first process is activated. There is however an instruction, equivalent to the FORK in the original PRAM, which allows the activation of new processes to execute any portion of the program. The difference resides only in the fact that, whereas in the PRAM case FORK activates a real processor, here it activates a process. The execution of a FORK instruction corresponds to a "branching" of the execution flow with the creation of parallel paths on the computation graph. Regarding the computation graph however we have seen that there is also the opposite need, precisely the need to join several paths at a single point. This represents the case when a given operation can be executed only when a number of other operations have been executed. To satisfy such a need we have the JOIN instruction. Executing "JOIN n" joins n independent instruction flows (processes) into a single flow. One can visualize this situation thinking that the first $n - 1$ processes executing JOIN end their activity, and that only the nth process can go on to the next instruction. The following example is a version of the program in Example 5.5, complete with the process activation scheme.

Example 5.7 (sum of $N = 2^n$ numbers)

```
for k ← 1 until N-1 do
    FORK init;
```
                                comment $N-1$ processes starting from the instruction with the label *init* are activated;

```
k ← 0;
init:  if (k mod 2) = 0 then E[K] ← 0
       else begin
           E[K] ← 1;
           JOIN n;
```
                                comment processes with odd k are used only to initialize flags;

```
       end;
       for i ← 2 step i until N do
           if (k mod i) = 0 then begin
               WAIT(E[k+i/2]);
               A[k] ← A[k]+A[k+i/2];
               SIGNAL(E[k]) end
           else JOIN n;
       JOIN n;
```
                                  comment only the process with $k = 0$ gets to this point;

...

It is clear that the FORK and JOIN mechanisms are also in need of an accurate use in order to arrive at determined and correct computations.

5.2 Interconnection structures

In this section we will discuss a few structures used to connect processing elements and implement a parallel machine. The type of parallel machine under investigation consists of processing units connected among themselves, and is, in general, without a common memory.

To study interconnecting structures it is useful to recall some notions on graphs. We can indeed establish a correspondence between the nodes of a graph and processing elements, as between arcs and connections. Thus, the

188   Parallel architectures

study of the topological properties of the structures is easier when referring to the corresponding graph properties. From our point of view, a parallel architecture can be seen as a set of processors interconnected by means of links transporting messages. In particular, when attention is focused on the interconnecting *pattern*, a graph naturally corresponds to the architecture, as mentioned above. (For a complete and detailed discussion of the characteristics of the main interconnecting networks see [WF84].)

While presenting different kinds of structures, we will sometimes refer to situations when "N senders" communicate with "N receivers", where sender and receiver can be processor and memory (or vice versa) or two processing elements.

The interconnecting structure is made of arcs and possibly of switching points. These points implement a given routing protocol. To access the structure, each unit has input and output hardware resources, called *communication ports* (see Section 3.6). The classification will be done on the basis of the behaviour of the structure complexity as a function of N, where the complexity itself is given by the number of communication ports each unit has, times the number of units. In the following, we will examine structures with complexity $O(N)$, $O(N \log N)$ and $O(N^2)$.

5.2.1 Structures with O(N) complexity

The architectural cost of these structures increases linearly with the number of interconnected units. The most widely used structure, and not only among those with a linear cost, is perhaps the asynchronous parallel bus (Figure 5.4). This is implemented as a series of one-bit parallel lines plus an arbitration mechanism. Only one unit at a time can access the bus to write to it, after asking and obtaining the assent of the arbitration mechanism. All units are, on the other hand, automatically allowed to read. Besides the low cost, the main advantage of a structure of this kind is the possibility of efficiently using, as a consequence of serialization of transmissions, the whole *transmission band*.[4] The main disadvantage is that, since a single resource is used by all units, even for low values of N (20 or 30) the bus becomes a bottleneck with a negative influence on system performance (one also says that it has a reduced *scaling capability*).

---

[4]The transmission band is the number of bits that can be transferred in a unit time.

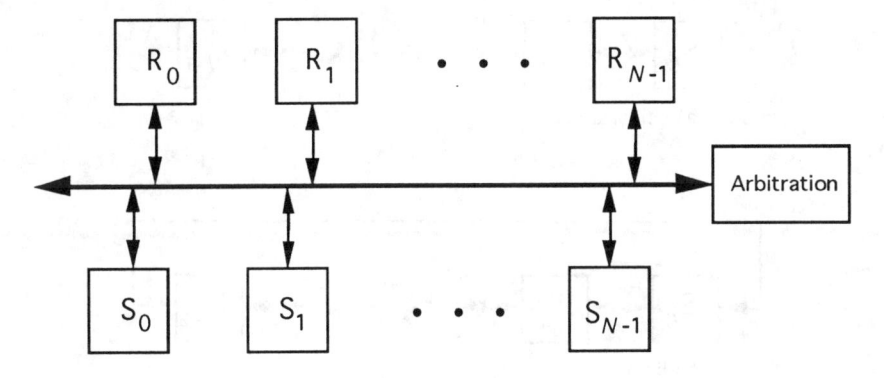

Figure 5.4 Bus type interconnection.

The bus is the only linear cost structure ensuring a "direct connectivity" between every possible sender and every possible receiver. By direct connectivity we mean that messages, once input by the sender into the interconnecting structure, exit the structure only at the interface with the receiving unit. Other structures with cost $O(N)$ do not have this property. In such cases a message, to reach its destination, must be input several times into the interconnecting structure by intermediate units, acting as routing units. Examples are unidimensional structures such as the linear array and the ring (Figure 5.5), and two-dimensional ones, such as the binary tree and the processor mesh (Figure 5.6). Every "node" in these structures consists of a processing element.

One-stage networks also have $O(N)$ complexity. These are networks where messages go, within the structure, through a single switching point. As in the case of unidimensional and two-dimensional structures, these networks also have a limited connectivity, and in general it is necessary for a message to cross the network several times before getting to its destination.

One of the most interesting examples of a one-stage network is the *perfect shuffle* (Figure 5.7(a)). Figure 5.7(b) shows the *interconnecting graph* of this network. Each node represents a *processing element*[5] (PE), while a connection between node i and node j indicates that PE_i can directly communicate with PE_j. The two different graphs refer to the two different functions which can be applied at the switching points (switches S in Figure 5.7(a)).

---

[5]By the term processing element or PE one denotes any unit able to receive messages, perform more or less complex computations, and store data.

190 Parallel architectures

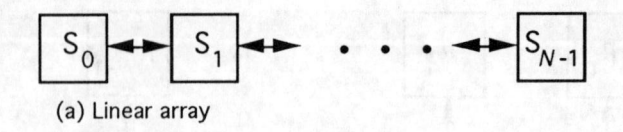

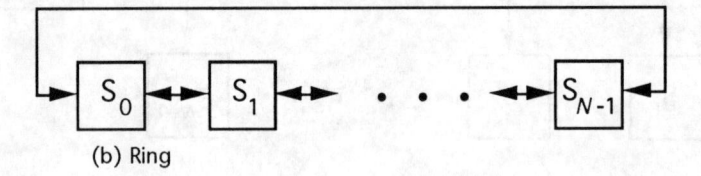

(a) Linear array

$S_0 \leftrightarrow S_1 \leftrightarrow \cdots \leftrightarrow S_{N-1}$

(b) Ring

Figure 5.5 One-dimensional structures.

(a) Binary tree

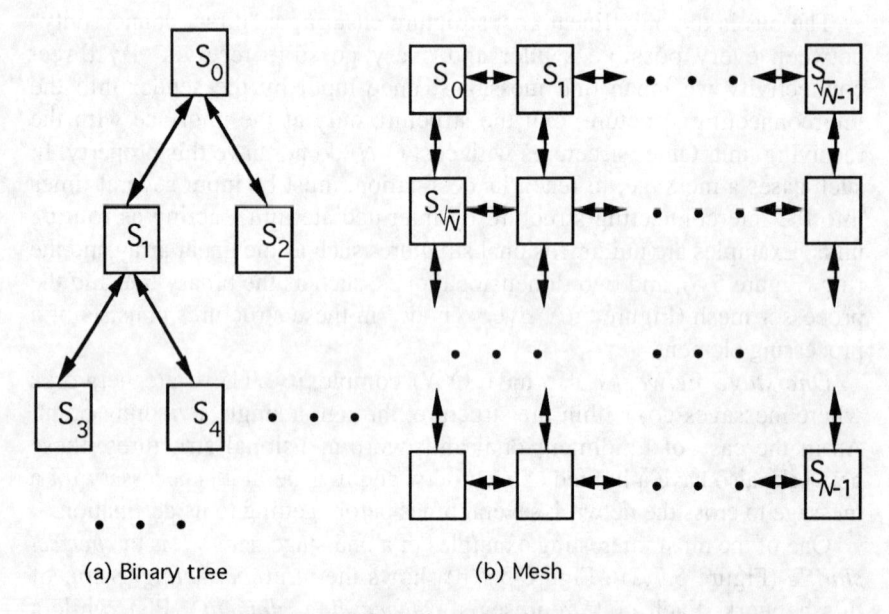

(b) Mesh

Figure 5.6 Two-dimensional structures.

In one-stage networks organized as a perfect shuffle one can obtain complete connectivity by sending a message log N times through the network, and by programming a different behaviour of the switching points each time. At each "step" (single transmission) an intermediate PE is given the task of storing and then forwarding (*store-and-forward*) the message. It is apparent that these operations have a cost in terms of time and memory space, and it is also clear that this is the price to be paid to have a low complexity network.

Interconnection structures **191**

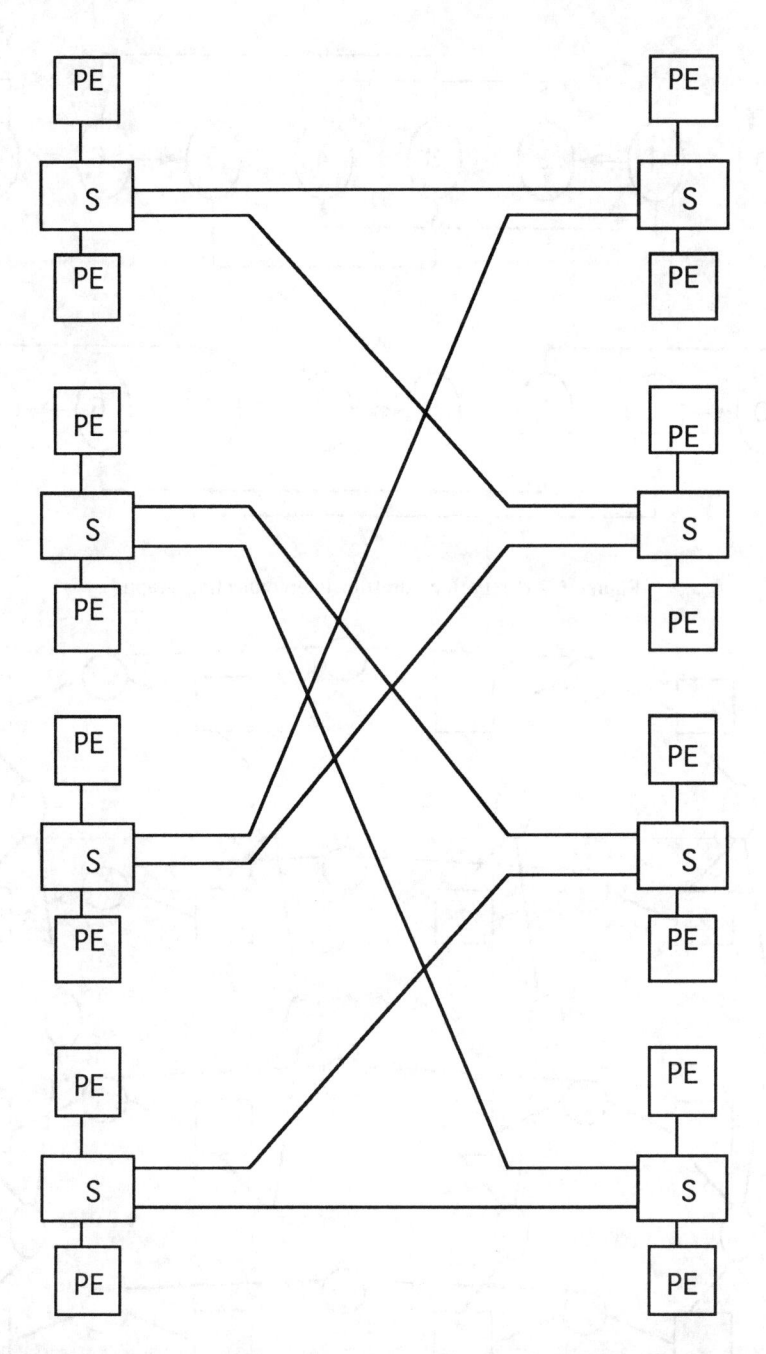

Figure 5.7 (a) Perfect shuffle: interconnection structure (S = switch; PE = processing element).

192 Parallel architectures

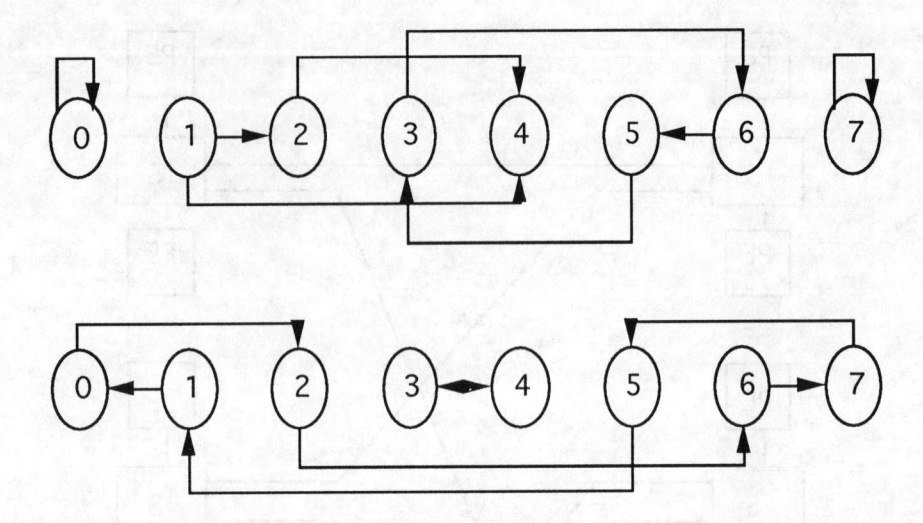

Figure 5.7 (b) Perfect shuffle: interconnecting graph.

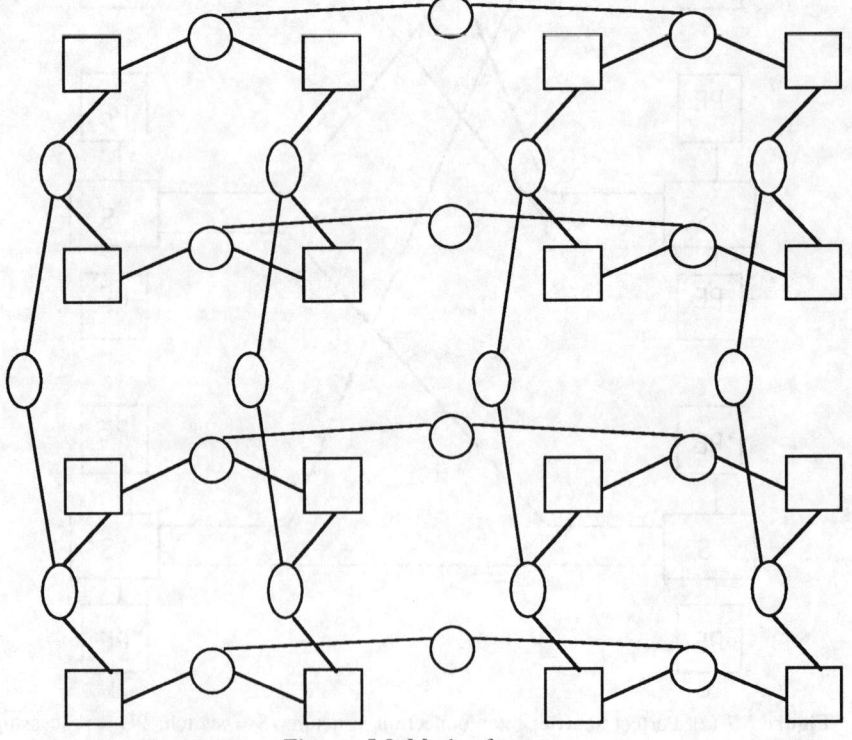

Figure 5.8 Mesh of trees.

Another linear complexity interconnecting network can be obtained by combining a mesh with a binary tree, in such a way as to have a mesh with trees connecting both the elements on a row and the elements on a column. Such a structure is called a *mesh of trees* (Figure 5.8).

5.2.2 Structures of O(Nlog N) complexity

These structures allow complete connectivity by means of log N levels (or stages). In fact, since every switching point defines two possible routing functions, log N levels are sufficient to reach every possible destination starting from every possible source (provided the connections between switching points are suitably chosen).

Figures 5.9 and 5.10 show two of the most common networks with log N stages proposed in the literature: *butterfly* (or *banyan*) *network* and *cube network* respectively. It is interesting to analyze the use of *routing tag* (information carried by the message) to control multistage networks. The simplest case is the cube network. Let $i = b_{k-1}b_{k-2} ... b_1b_0$ be the address of the sender PE and $j = b'_{K-1}b'_{k-2}...b'_1b'_0$ the address of the destination PE. The sender PE forms the exclusive OR of i and j seen as bit sequences, that is, it computes $c_t = b_t \oplus b'_t$, $t = 0, ..., k-1$. c_t is the routing tag for the stage $k - t + 1$, and if $c_t = 0$, the inputs are sent directly to the corresponding outputs. Conversely, if $c_t = 1$, inputs and outputs are exchanged.

Figure 5.11 shows the *Boolean cube* (or *hypercube*), which is another especially important interconnection structure. In this case the complexity $O(N \log N)$ refers to the number of connections in the processor network.

5.2.3 Structures of O(N²) complexity

The structure with $O(N^2)$ complexity ensures complete interconnections with a better behaviour from the point of view of transfer time. The crossbar (Figure 5.12) is the classical example. It consists of $2N$ buses and N^2 switching points. Note that every path crosses at most a single switching point, and that in the best case the crossbar allows the parallel transmission of N messages. Owing to its high cost, the crossbar can be used only for low values of N (about 20 or 30).

Among the structures with $O(N^2)$ complexity we also include the network without switching points, which provides a dedicated connection between each pair of units (Figure 5.13). In this structure it is the number of connections itself, and therefore the number of communication ports in each unit, which grows quadratically. A redundancy of this kind cannot be justified, since the increase in performance of a system with p processors with respect to a single processor system is (at most) linear.

194 Parallel architectures

Recall that this "solution" was used in the abstract model of a network with variable (or computed) interconnecting structure introduced in Chapter 3.

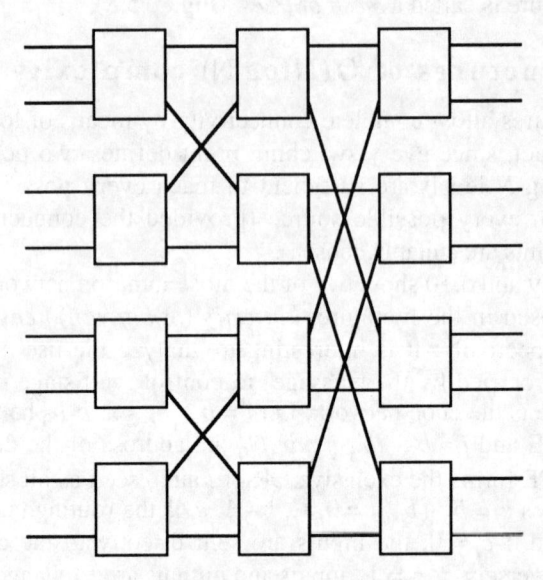

Figure 5.9 Butterfly.

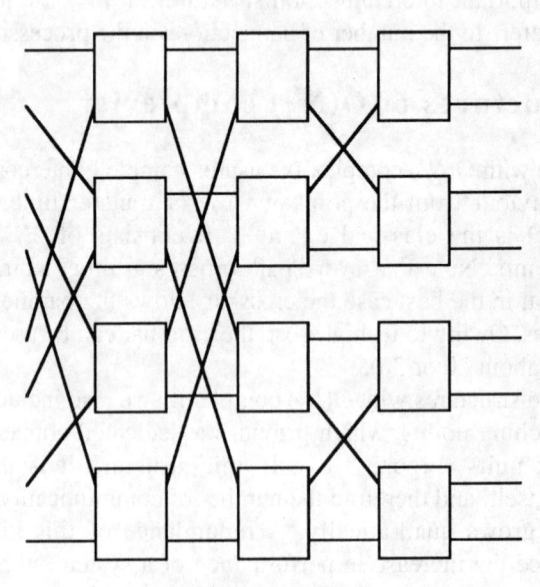

Figure 5.10 Cube.

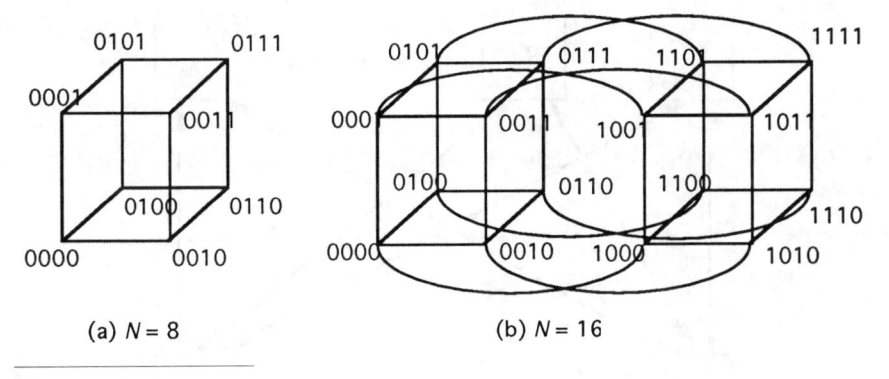

Figure 5.11 Hypercube.

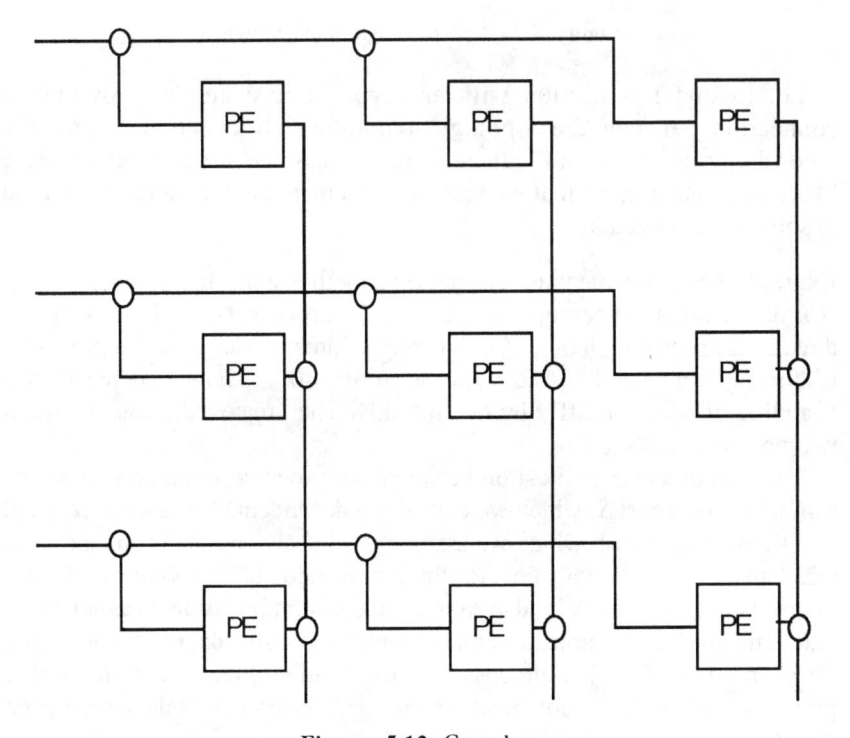

Figure 5.12 Crossbar.

As we shall see in Chapter 6, the study of specific problems and algorithms will determine which structure allows the execution of efficient computations.

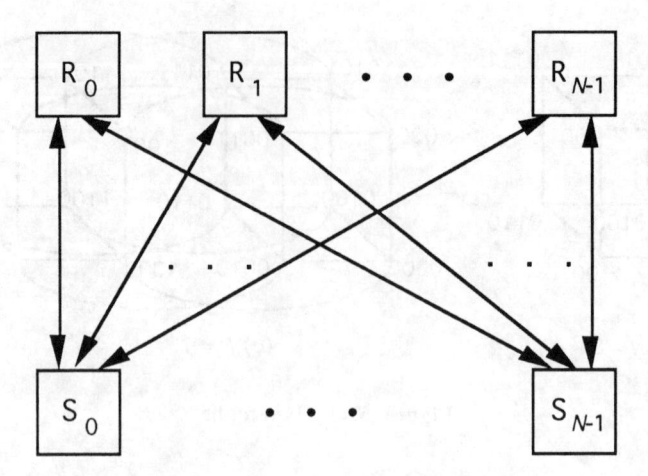

Figure 5.13 Complete interconnection.

Let us end this section with an algorithmic example, providing a connection between the topological notions introduced here and the algorithmic problems of Chapter 4. Let us assume we have available a hypercube, and assume that we want to implement on it an iterative method to solve linear systems.

Example 5.8 (Implementation of iterative methods on a hypercube)
A k-dimensional hypercube consists of 2^k processors, each of which is directly connected with other k processors. More precisely, the nth processor is directly connected to the mth one, $n, m = 0, ..., 2^k - 1$, if the binary encoding of n and m differ by one bit only. The greatest distance between two processors is k.

The cost of a communication between two processors can be seen as the sum of a *set-up* cost S (which we consider independent of message size) and of a transmission cost, which we assume is a linear function of the message size l, that is $T*l$. The total cost for the transmission of a message of size l is then given by $S + Tl$. We also assume the communication channel to be bidirectional, and that transmission can simultaneously take place both ways. Thus, in a time $S + Tl$, a processor can send to and receive from another processor, to which it is not directly connected, an amount of data not greater than l.

The main problem arising in the implementation of iterative methods to solve linear systems is that of determining mapping schemes between data (elements of the coefficient matrix) and processors, in such a way as to minimize the amount of data to be transferred.

In the following, we will analyze the conjugated gradient method ([GVL83]).

Consider the following formulation of the conjugated gradient method to solve a linear system $Ax = b$.

Let $i = 0$ and $r_0 = b - Ax_0$ (where x_0 is arbitrary).
   while r_i is not sufficiently small **do**
      1.   Compute $q_i = Ap_i$.
      2.   Compute $p_i^T q_i$ and $\alpha_i = r_i^T r_i / p_i^T q_i$.
      3.   Compute $r_{i+1} = r_i - \alpha_i q_i$.
      4.   Compute $x_{i+1} = x_i + \alpha_i p_i$.
      5.   Compute $r_{i+1}^T r_{i+1}$ and $\beta_i = r_{i+1}^T r_{i+1} / r_i^T r_i$.
      6.   Compute $p_{i+1} = r_{i+1} + \beta_i p_i$.
      7.   Let $i = i + 1$.
end.

Assume A is a piecewise matrix. Assume further we have a two-dimensional hypercube (Figure 5.14), and assume that the data-processor mapping is the one shown in Figure 5.15. We want to analyze the cost of one iteration in terms of computation and communication steps.

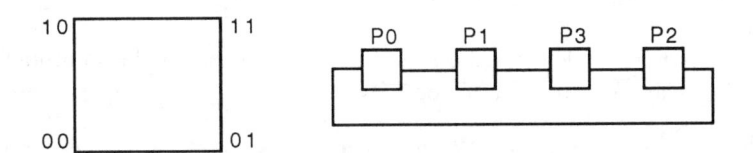

Figure 5.14 Two-dimensional hypercube.

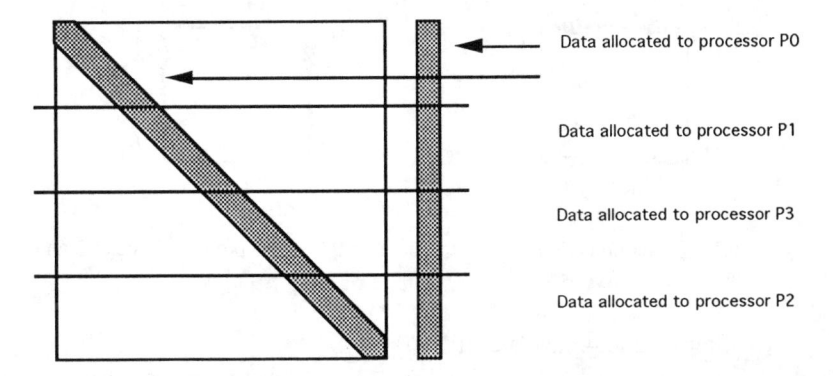

Figure 5.15 Data-processor mapping.

198 Parallel architectures

Let us assume that the size n of the matrix is a multiple of 4. Let $m = n/4$ and let $\mathbf{v}^{(k)}$, $k = 0, 1, 2, 3$, the vector $[v_{km+1}, v_{km+2}, ..., v_{(k+1)m}]^T$. Further, let z be the average value of the number of non-vanishing elements in each row of matrix A.

- At step 1 processor k must receive data (that is, the elements of vector $\mathbf{p}^{(k-1)}$ and/or vector $\mathbf{p}^{(k+1)}$) from one of the adjacent processors in order to compute $\mathbf{q}_i^{(k)}$. Thus, step 1 costs $(2z-1)m$ arithmetic operations and at most two communications. Owing to synchronization, the cost C_1 of step 1 is:

$$C_1 = \max_k \{Tl_k\} + 2S + (2z-1)mT_{op}$$

where l_k is the amount of data exchanged by processor k, while T_{op} denotes the cost (in time units) of an arithmetic operation.

- At step 2, processor k computes the scalar product $\mathbf{p}_i^{(k)T}\mathbf{q}_i^{(k)}$ executing $2m - 1$ arithmetic operations. Without communications, however, it is not possible to calculate the global constant α_i. Following the communication scheme shown in Figure 5.16, each processor can compute $\mathbf{p}_i^T\mathbf{q}_i$ in a time:

$$T_{exe} = 2(S + Tw + 1)$$

where w denotes the size of the *data* used in the computation. Therefore the cost of step 2 is:

$$C_2 = (2m - 1)T_{op} + T_{exe} + T_{op}$$

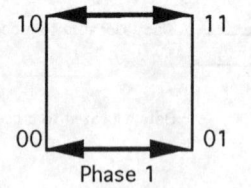

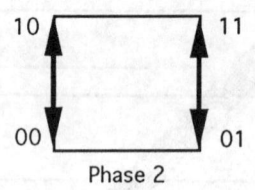

Figure 5.16 Communication pattern to compute the scalar product in steps 2 and 5.

- Steps 3 and 4 have a cost:

$$C_3 = C_4 = 2mT_{op}$$

- Step 5 has the same cost as step 2:

$$C_5 = (2m - 1)T_{op} + T_{exe} + T_{op}$$

- Step 6 has the same cost as steps 3 and 4:

$$C_6 = 2mT_{op}$$

The total cost of one iteration of the conjugate gradient method on a bidimensional hypercube (4 processors) is then:

$$C^{(4)} = \max_{k} \{Tl_k\} + (2z + 9)mT_{op} + 2S + 2T_{ex}$$

It is easy to verify that the sequential cost of one iteration is:

$$C^{(1)} = (8z + 36)mT_{op}$$

and this leads to a speed-up of:

$$S_4(n) = \frac{(2z + 9)nT_{op}}{(z/2 + 9/4)nT_{op} + \max_{k}(Tl_k) + 2S + 2T_{exc}}$$

In [AOES88] there are a few experiments performed on an hypercube iPSC/2 showing that the speed-up $S_4(n)$ is not less than 3.

5.3 Shared memory architectures

By the term *multiprocessor* we mean a parallel architecture with several processors and one or more memory units. From the points of view of data access and communication between processors we can distinguish two separate models, one with *common* or *shared memory* and one with *distributed memory*, corresponding to the abstract PRAM (Section 3.4) and processor network (Sections 3.6 and 5.2) models. In the case of a common memory multiprocessor system, one considers a bidirectional interconnecting structure with p processors and m memory models (Figure 5.17). On the other hand, in the case of a distributed memory multiprocessor, each of the p processing elements (the term processing element here denotes a processor with its local memory) is connected to an input and an output of a unidirectional interconnecting structure (Figure 5.18).

In this section we will focus our attention on some topics related to computing environments with a common memory. Regarding computations, the ideal situation is when the execution time of an algorithm on a machine with p processors is p times less than the time required by the best sequential algorithm for the same problem. Even in the case when there is the real possibility of dividing the computation in at least p different activities, such an ideal situation still cannot be reached due to *conflicts* between processors for the use of common resources. We can distinguish at least four kinds of such resources [RET86]: the interconnecting structure, single paths within the structure, memory modules, and single locations within the memory modules. Conflicts for the use of the last two resources take place only in the case of common memory multiprocessors.

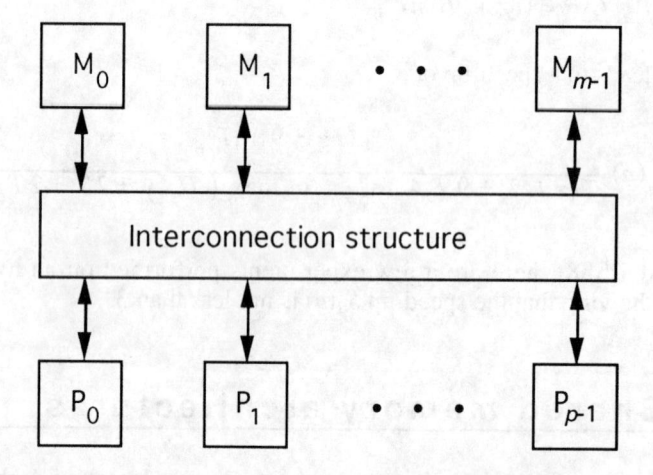

Figure 5.17 Shared memory multiprocessor.

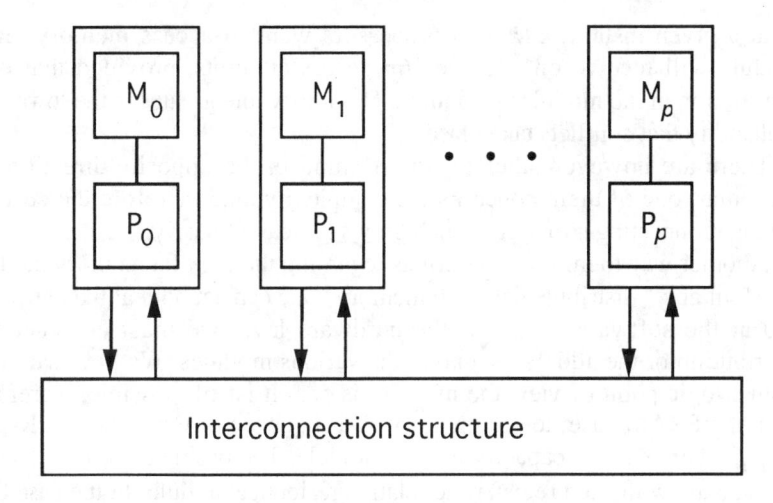

Figure 5.18 Multiprocessor without shared memory.

The fact that, in shared memory systems, the interconnecting structure is bidimensional means that data can move both ways on the logic connection linking processor P_i and memory module M_j. From a logic point of view the memory is one, even if it is divided into modules. The partition is needed to allow several processors simultaneous access to memory locations residing in different modules. This problem was all but ignored in the abstract PRAM model.[6] In a real case a memory model can be subsequently accessed only by one processor. If more than one processor wants to write to or read from locations, even if different, contained in the same module, accesses need to be serialized. This means that an order must be established according to which requests are satisfied one after the other and not in a parallel way. In the case of a memory implemented as a single module one then has conflicts every time more than one processor needs to write to or read from memory a datum. It is apparent that a situation of this kind can have serious implications on the total time necessary to complete a computation, as soon as it is assumed that writing to or reading from a memory location requires a non-zero time.

The need to minimize this kind of conflict pushes towards a division of the memory into a number of modules. Roughly speaking, we can say that

---

[6]In the PRAM case one considered only conflicts regarding the access to a particular memory location and not to the memory as a whole. It is clear that these two possibilities are coincident only in the (unlikely) hypothesis that each memory location is implemented as a stand-alone module.

if, at a given instant, k ($k \leq p$) processors want to access memory, each module will receive on average k/m access attempts, provided that data distribution in the modules is random. Obviously, the greater m the lower the probability that conflicts take place.

There are however other factors pushing in the opposite direction. In particular, one of them concerns the complexity, and therefore the cost, of the interconnecting structure, which grows (at least) linearly.

Another way to minimize conflicts regarding the use of a memory module is to "suitably" distribute data into memory. One can act both at the hardware and at the software level. At the hardware level we must consider the distribution of the addresses among the various modules. We have said that from a logic point of view the memory is one. It is still possible, therefore, as in the PRAM case, to see memory as a sequence of registers: $R_0, R_1, ..., R_{cm-1}$, where c is the capacity of each module. The problem we now face is to associate with each register the relative residence module. In the case of a *sequential* organization the registers $R_0, ..., R_{c-1}$ are assigned to module M_0, the registers $R_c, ..., R_{2c-1}$ are assigned to module M_1, and so on. If then i is an integer such that $0 \leq i < cm$, and if we let $r = \lfloor i/m \rfloor$ and $s = (i \bmod m)$, a reference to register R_i translates into a reference to the sth register of M_r. In the case of an *interconnected* organization the registers $R_0, R_m, ..., R_{(c-1)m}$ are assigned to module M_0, the registers $R_1, R_{m+1}, ..., R_{(c-1)m+1}$ are assigned to module M_1, and so on. If i, r and s are as in the previous case, a reference to register R_i translates into a reference to the rth register of M_s.

At the software level one considers a data distribution in the various modules such that the maximum possible load balance is ensured during program execution. It is clear that such a distribution must be done automatically, both in an implicit way (by the compiler) and in an explicit way (by means of a call to suitable allocation subprograms). Obviously, at the software level one must take into account the hardware organization, whether sequential or interconnected.

Example 5.9
Consider the computation of the infinite norm of an $n \times n$ matrix, that is:

$$\|A\| = \max_{0 \leq i \leq n-1} \sum_{j=0}^{n-1} |a_{ij}|$$

Computing this quantity requires first the computation of the n sums $S_i = |a_{i0}| + ... + |a_{i,n-1}|$, $i = 0, ..., n-1$, then the determination of the maximum of $\{S_i\}_{i=0, ..., n-1}$. If we have at least n processors we can assign the computation of S_i to processor P_i, $i = 0, ..., n - 1$. In this case, processor P_i must be able to access all, and only, the elements of the ith row. Assuming, for the sake of simplicity, that the dimension of the matrix coincides with the number m of memory modules, one can proceed in the following way. With an interconnected organization the matrix is stored by columns, that is, its elements are written to consecutive locations changing the row index faster than the column index. If b is the *storage base* of the matrix in memory, register R_b will contain the element a_{00}, R_{b+1} will contain a_{10}, and so on. It is easy to realize that in this way each row will be stored in a different module. Namely, the ith row in module $(i + b)$ mod m. A similar result can be obtained, in the case of a sequential organization, by storing the matrix by rows. Note finally that the computation of $\|A\|$ as the maximum of $\{S_i\}_{i=0, ..., n-1}$ can be done in $\lceil \log n \rceil$ steps by means of the usual tree scheme (see Section 3.4).

The problem of data distribution minimizing conflicts for access to the same memory model is not as simple as the previous example may lead one to think. Indeed, the same program could originate computations behaving differently with respect to data access. In these cases, "easy" solutions, as in the example, are not so immediate. In Example 5.9, if one needed to calculate, besides the values S_i, also quantities depending on values contained in the same matrix column, each processor would have to access every memory module.

In the course of a parallel computation, the results produced by some processors must be liable to further processing by some other processor. If there exists a common memory, it is natural to think of it as a means of communication, as well as a "storage area" for data: a processor "deposits" a message in a well-defined memory area, and, subsequently, another processor "withdraws" it. Recall, in this respect, the examples of PRAM computations seen in Chapter 3. The communication, or *direct interaction*, problem between processors then becomes a problem of *indirect interaction* through the use of a common resource. The main problem here is the need for data level synchronization. This means ensuring that the message destination process does not read the contents of the area set aside to contain the message itself before the latter has been completely written.[7] Obviously, in the programming language made available by the firmware architecture there are mechanisms for obtaining a predetermined behaviour. Among the

---

[7] If a message is n bits long, and if each write to memory can involve at most n' bits, the complete message writing requires $\lceil n/n' \rceil$ accesses. It is necessary to ensure that a read operation is not inserted between these write operations by the destination processor, which could then access inconsistent information.

most common are the WAIT and SIGNAL primitives, which have been introduced in Section 5.1. Here we only want to stress the fact that the use of these mechanisms (that is, the programmer's view of the memory resource) itself can be a source of serious mistakes. As we have already seen, in the case of asynchronous programming, the problem of process level synchronization adds to the problem of data level synchronization.

5.4 Message-passing architectures

In the distributed memory model each of the p computing elements (PE) is connected to a unidirectional interconnection structure (Figure 5.18). This means that the fact that PE_i is able to send messages to PE_j does not imply that PE_j is able to send messages to PE_i. If it is necessary for two computing elements to communicate between themselves one must then explicitly provide two unidirectional connections.

Each processor has exclusive access to its own local memory. One assumes that most of the read/write requests by the processor itself are addressed to this memory. This obviously depends on the way data have been subdivided, and, at the end, on the particular type of computation. Non-local data access always requires communication with the PE where data are stored. In this case then data distribution can influence *data routing* through the communication structure, and therefore the number of conflicts on path usage. Obviously, there are no conflicts on memory usage.

The above discussion justifies the term *message-passing architecture* which is often used for distributed memory multiprocessors. In fact, communication between processors takes place by passing an information structure, called a message, where information such as sender processor, destination processor and the contents of the message itself is specified. At the firmware level programming language we will then have communication primitives, in the form of instructions such as SEND and RECEIVE (see Section 5.1), which ensure a correct message transmission. It is apparent that the presence of the above mechanisms eliminates, in distributed memory architectures, the data level synchronization problem. We have however already seen that in the case of asynchronous programming a process level synchronization could still be necessary.

Despite the fact that it is widely accepted that common memory provides the most natural solution to parallel algorithm communication problems, there are enough reasons to consider the distributed memory model more promising for machines with massive parallelism. Two of these reasons have a physical character, while the third is a logic one.

- When the number of processors increases the common memory becomes a bottleneck for the whole system due to the large number of accesses and consequent possible conflicts.
- The complexity of an interconnecting structure able to handle the communication load between processors and memory weighs considerably on the cost of the whole system when the number of such units grows. Further, the structure is always a source of possible conflicts between processors.
- The presence of synchronization problems both on the data and on the process level makes the programming of a common memory multiprocessor "dirtier" from a formal point of view and generates more errors.

In the case of the distributed memory architecture the conflicts for memory access disappear, while interconnecting structure usage conflicts persist. These conflicts increase with the number of interprocessor communications. However, if the interconnection structure reflects the structure of the computations the routing load can be kept at reasonably low levels. This last observation legitimates the hypothesis that the common memory architecture could be better suited to general computations, while particular distributed memory architectures (according to the interconnection topology) could be more efficient for specific computations. In the following chapter we will see how VLSI technology enables one to integrate, in a single system, interconnection networks reflecting the topology of the computation graphs.

Bibliographic notes

An introduction to parallel architectures is given in *Computer Architecture and Parallel Processing* [HWB84], which also contains a rather complete bibliography.

Several noteworthy results on parallel architectures, both from a computational and from a topological point of view, have been presented in scientific papers in the periodical *IEEE Transactions on Computers*, starting since 1970.

For a deeper discussion of the notions treated in Section 5.2, we suggest reading the review contained in the book *Interconnection Networks for Parallel and Distributed Processing* by Wu and Feng, published by the IEEE Computer Society (1984) [WF84], in addition to [GIL87]. To study interconnection structures that we have left out of our discussion see [PV81], [STO71], [STO75] and [Si79]. A hypercube emulator, the *cube connected cycles*, described in [PV81] is particularly important.

Regarding relations between architectures and computational models see [Gol78] and [Gol82]. Finally, in the book *Designing Efficient Algorithms for Parallel Computers* by Michael Quinn [Qu87], one can find many parallel algorithms, with special attention being given to the more concrete aspects and to the connection with architectures.

Problems

5.1 If a computation requires more communications than operations, what are the advantages and disadvantages in using SIMD vs MIMD architectures?

5.2 On the basis of what criteria can one decide to use a SIMD or a MIMD programming style?

5.3 List the main disadvantages of the local memory architectures.

5.4 How much time does a hypercube take to simulate a complete interconnection?

5.5 Classify the interconnection networks introduced in this chapter on the basis of their "power". (Hint. Use simulation arguments.)

5.6 Execute the mergesort algorithm on a hypercube, CCC and shuffle. Analyze the performance.

5.7 Using the technique of Example 5.8, compute the nth power of an $n \times n$ matrix on the hypercube.

5.8 State which properties of a common memory multiprocessor (as described in Chapter 5) are ignored in the PRAM model.

5.9 Quantify the effect of synchronization primitives on algorithm performance. (Try to obtain worst case and average case measures.)

Chapter 6
VLSI Algorithms and Architectures

During recent years a technology has been developed to produce circuits on a semiconductor (silicon) lattice. These circuits consist of *integrating* hundreds of thousands of transistors. This technology is called Very Large Scale Integration, usually indicated VLSI. Integrated circuits have changed the situation where the design of electronic circuits neglected the cost of links with respect to the cost of the switching elements. Most of the space within integrated circuits is occupied by the switching elements which take most of the time, in contrast to what happened before.

A new theoretical field has grown, dedicated to the research of computing models outlining the characteristics of VLSI architectures. These computing models cannot neglect the costs of links, as was the case with models highlighting the crucial resources in conventional computers. In the latter case the physical limitations imposed by the links were minor with respect to the logical limitations. It follows that it is necessary to go from models mainly devoted to operations (such as TM, RAM or PRAM), to models which concentrate on the need for distributing devices in space and for interconnecting these devices.

It is apparent that there has been a passage from a situation (based on the concept and technology of the conventional computer) where the logical aspect was fundamental to a situation where the distribution and geometry of links have the same importance. Sutherland and Mead, in a paper published in 1977 [SM77], anticipated that the designers of computing models for VLSI had to add a geometric regularity to the logic regularity that had already been reached and evaluated.

One of the main objectives of this chapter is to describe some kind of "computational theory" for VLSI. This description involves several stages: after treating (in a simple way) the characteristics of silicon integration, we will describe computing models and criteria well suited for analyzing the algorithmic and circuital complexity, which will later be used in evaluating VLSI performance.

208 VLSI algorithms and architectures

From a computational point of view the main characteristic of VLSI is that there is no difference between software (algorithms) and hardware (circuits). For each algorithm there is a corresponding circuit implementing it. From a practical point of view this allows one to integrate complex functionalities directly on silicon. From a theoretical point of view it provides a natural hardware measure, i.e. the circuit area. Since this measure is adequate for circuit costs, it will appear in every complexity analysis.

We will see in the following that a formal way to analyze a VLSI architecture is to associate it to a graph. Its nodes and arcs correspond to computing elements and interconnections respectively. A focal point is then to design a *layout* with minimum area for such a graph.[1] For this purpose many tools of analysis will be imported from the part of graph theory concerned with *embedding* graphs in the plane. In the following sections we will determine lower and upper bounds on circuit area. To this end we will sometimes use results related to the notions of *bisection*, *separator*, *wire area* and *crossing number* [Th79], [Th80], [Le81], [Ya81], [Bi84].

The algorithms–circuits correlation mentioned above has some important consequences. Above all, given a computational problem and an algorithm to solve it, it is possible to associate to this algorithm a graph (i.e. its computation graph). One must look for its minimum area layout, obtaining an efficient VLSI algorithm in terms of occupied area.

This is the reason why VLSI is often considered a powerful tool for obtaining algorithms directly on hardware.

There are two main approaches to VLSI computational complexity. On the one hand, there is the search for special–purpose architectures for the efficient (possibly optimum) solution to a given problem. On the other hand, there is the search for versatile architectures for solving several computational problems with a good, even if non–optimum, performance. An example of versatility is the shuffle–exchange graph, which is amenable for implementing FFT algorithms, fast–sorting algorithms and algorithms for polynomial computations [Le81].

In this chapter we first make some preliminary observations on the technology (Section 6.1), then we describe the chosen computing model (Section 6.2). Section 6.3 concentrates on VLSI complexity measures, while Section 6.4 considers the complexity of some arithmetic and numerical problems.

6.1 Introduction to VLSI technology

The microelectronic revolution began with the construction of the integrated circuit on silicon. When integration was not yet possible, each single component of an electronic circuit was built separately and linked to the other components in a second step. It is now possible to build on a *semiconductor*

---

[1] A graph layout is the disposition, based on appropriate rules, of the graph on a grid.

(silicon) plate both the active components (transistors and diodes) and the passive components (resistors, capacitors and inductors). This allows one to obtain a complete electronic circuit, which can currently contain up to a million transistors on less than a centimetre of square plate. For this reason, very large scale integration (VLSI) introduced a revolution in the design of electronic circuits. In recent decades the *degree of integration*, i.e. the number of integrated components on the same circuit, has roughly doubled every two years. At the same time the costs have been decreased, so that a great number of components is now accessible.

The evolution from a single transistor to modern VLSI has been a continuous process connected to advances in technology. At the microelectronic level, by technology we mean the materials used to build the layers and the techniques used to place them one on top of the other. The silicon basis on which the circuit is built is called a *wafer*. When the wafer has completed the passage through the building stages, it is sectioned in chips, each one containing a complete circuit.

Clearly, the main circuit component is silicon. Hence the development of integration techniques is dependent on the development of technologies for building electronic devices on a silicon grid. The easiest component to be integrated on silicon is the transistor. The abundance of transistors in electronic devices makes up for the difficulties which are present in building passive components on silicon. The property which makes transistors necessary on an electronic circuit is their gain capacity, that is, the ability to raise the power of a low input signal.

In order to investigate how transistors (and other active and passive components) are made, we must first consider the nature of a semiconductor. A silicon atom has four valence electrons which, in a solid silicon crystal, are shared among adjacent atoms in such a way as to surround each atom with eight electrons. All electrons are bound, therefore silicon is not a good conductor. To make, starting from silicon, a semiconductor crystal it is necessary to perform a *doping* process consisting of the introduction of impurity atoms within the crystal.

Typically, silicon is doped with phosphorous and/or boron. The atoms of these elements have five and three valence electrons respectively. When phosphorous is used to dope a silicon crystal, some electrons, one for each impurity atom, are not part of an interatomic binding and are therefore able to move under the influence of an external potential difference. Conversely, when the impurity is boron, there is a lack rather than an abundance of electrons, which is visualized by saying that there is one "hole" for each boron atom. Holes also move under the influence of an external potential difference. Phosphorous (or another pentavalent element) doped silicon is called an *n–type semiconductor* (where *n* stands for negative, indicating the negative electric charge of the electrons). Boron (or another trivalent element) doped silicon is called a *p–type semiconductor* (where *p* stands for positive, indicating the positive electric charge of the holes).

210 VLSI algorithms and architectures

The construction of the main circuit elements (diodes and transistors) is done with crystals doped in a different way. The basic electric phenomena take place in a region, called the *p–n junction*, which separates a *p*-type crystal from an *n*-type one.

The simplest semiconductor device is the diode (Figure 6.1(a)) which has the property of transmitting a signal with a given polarity while rejecting a signal with the opposite one. In other words, the diode allows the passage of a large current in only one direction. From Figure 6.1(a) it is clear that the diode is obtained from the interaction of only two doped crystals.

Since the diode does not provide a gain, the basic active element is the transistor. This device can be obtained by adding, with respect to the drawing shown in Figure 6.1(a), a third suitably doped crystal. By adjusting the size of the doped regions, one can confer a gain (amplification) capability to the transistor.

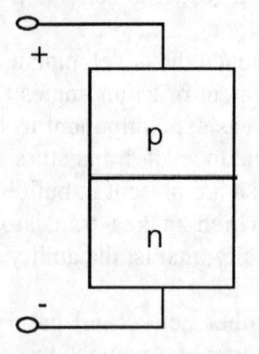

Figure 6.1(a) The diode. 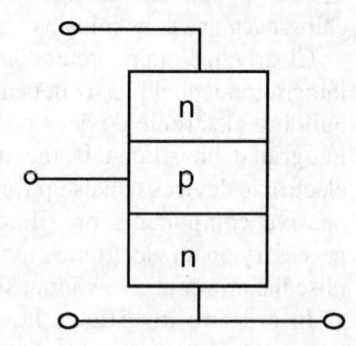   **Figure 6.1(b)** The transistor.

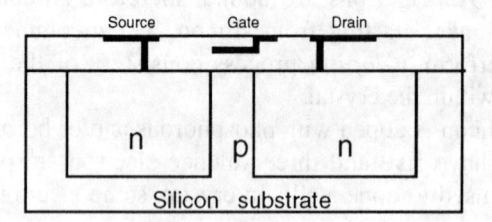

Figure 6.1(c) The nMOS transistor.

The transistor usually employed in microelectronics is the field–effect metal–oxide–semiconductor one (called MOSFET, from Metal Oxide Semiconductor Field–Effect Transistor). In the lattice of a doped semiconductor, there are two opposite doped regions, called the *source* and the *drain*, while a metal electrode, called the *gate*, is isolated from the lattice by means of a layer of silicon dioxide (Figure 6.1(c)). The transistor can control the electrical resistance between the source and the drain, by acting on the voltage applied to the gate. Two kinds of transistor can be obtained, nMOS and pMOS, according to the relative positions of the doped regions. Two types of technology, called NMOS and PMOS, correspond to these possibilities.

Using the semiconductor technology one can integrate on a single silicon crystal a large number of electronic circuits. The crystal itself, after doping, is cut in circular slices (wafer), on the surface of which integrated circuits are built. Each circuit occupies a small area (chip). At the end of the fabricating process the wafer is cut into chips.

The NMOS technology consists of three layers (*diffusion, polysilicon* and *metal*) where connections and the active and passive elements are built. Physically, the dopant is first deposited and diffused at a temperature of about 1000 °C (this creates the diffusion layer). A layer of polysilicon, which is a moderately conductive material, is deposited next, and finally a metal layer (typically aluminium).

For the sake of completeness we recall that the usually preferred technology is the CMOS (Complementary MOS) which overcomes some of the problems arising with NMOS, but at the price of a greater complexity. The CMOS technology includes nMOS and pMOS transistors on the same chip.

A particularly important technological aspect is the limiting value λ of the dimensions of some components (the value of λ depends on the technology being used). This limitation is imposed by electric considerations and it also reflects on limiting values for the distance between connections (cables). In NMOS design we can identify a few *general laws* (or design rules), which will be useful in the abstraction process leading, in the next section, to the definition of VLSI computational models. Let us now see in the following a few of these laws (P, M, D stand for polysilicon, metal and diffusion respectively).

Minimum width
- the connections of P and D must have a width $\geq 2\lambda$;
- the connections of M must have a width $\geq 3\lambda$;

Minimum separation
- the connections of P must be separated by a distance $\geq 2\lambda$;
- the connections D and M must be separated by a distance $\geq 3\lambda$;
- the connections between P and D, which do not form a transistor, must be separated by a distance $\geq \lambda$; (there are no constraints on the separation between the connections of M and one of the other two types).

Transistor formation
- if the connections of P and D must form a transistor, then P must extend at least 2λ over D; (if this were not the case, a displacement

212   VLSI algorithms and architectures

of λ both of D and of P could leave a small portion of D not covered by P, and therefore the current flow might never stop).

The laws described above are evidenced in Figures 6.2 and 6.3.

The nMOS transistor can be used in the fabrication of an *inverting* circuit, which in turn can be used to construct circuits implementing logically complete Boolean connections (NAND and NOR). The area used to integrate the Boolean operations on a silicon wafer is therefore proportional to the area of the transistor. A complete and detailed discussion of the fabrication of an inverting circuit, and of the construction, starting from this, of more complex devices, can be found in the classic textbook *Introduction to VLSI Systems* [MC80].

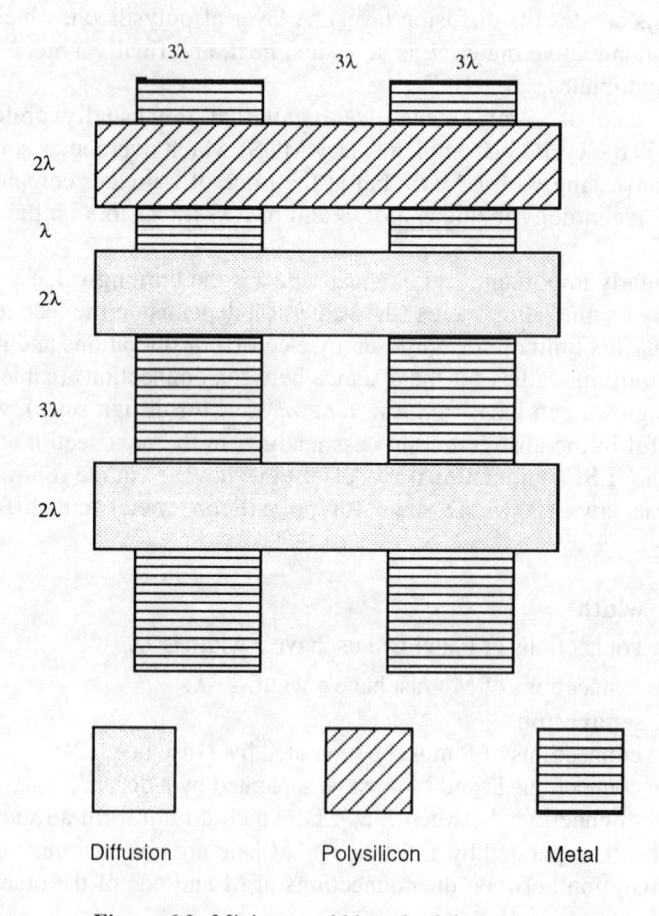

Figure 6.2 Minimum width and minimum separation.

Introduction to VLSI technology 213

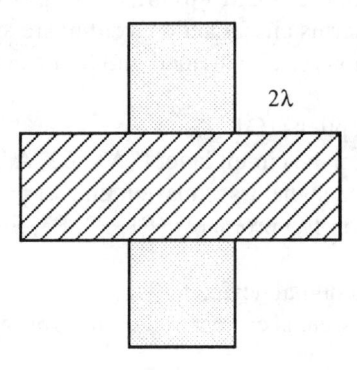

Figure 6.3 Transistor formation.

VLSI components made of several hundred thousand transistors are now commercially available. In this new context, the traditional measures used to evaluate the hardware complexity do not still seem to be adequate. As we have seen in Chapters 2 and 3, such measures stress the computational aspects more than the geometric and topological ones related to the disposition and the connection of devices. In the case of VLSI circuits, the large number of components requires consideration of the cost of data transmission among the various processing elements and of their disposition to be included in any architecture evaluation.

The abstraction process which will lead us to formalize a VLSI computational model will therefore have to take into account both the components needed for processing and those needed for connections.

6.2 VLSI computing models

The main characteristic of the approach discussed below consists of considering a VLSI circuit as a graph layout, where nodes correspond to computing elements, and arcs correspond to wires. More precisely, arcs simulate signal transmission between PEs, whereas nodes simulate dedicated PEs for both computation and input/output ports. This concept will become clearer in the following. Let us now start the abstraction process leading from the laws of the previous section to the definition of a model. For more details see [Th79], [Th80], [AA80], [AUY83], [BPP81], [CM81], [BG82], [Ya81].

As is usually the case for any mathematical object, a VLSI model must compromise between simplicity (useful for the mathematical investigation) and adherence to reality (needed to make reliable predictions or to draw correct conclusions).

We shall assume that a VLSI circuit is the physical fulfilment of an algorithm, so that the terms circuit and algorithm are synonymous from now on. The other assumptions can be divided into four categories:

- general assumptions (G1);
- assumptions on the circuit layout (A1, A2, A3, A4, A5);
- assumptions on signal transmission time (T1);
- assumptions on I/O protocols (P1, P2, P3, P4, P5, P6).

G1. A VLSI circuit is a digital device.

We assume that all signals correspond to an appropriate digital coding of the information.

Before introducing the assumptions on circuit layout, it is appropriate to consider the design rules of Section 6.1. These rules are due to the fact that, even though the layout of a VLSI circuit must have a small area, this area cannot be arbitrarily small due to technologic limitations on wire or transistor distances and dimensions. Violating this limitation implies a possible incorrect behaviour of the circuit.

A1. Wires have a minimum width λ.

For the sake of generality we shall consider λ as a parameter depending on the technology.

A2. Transistors have a minimum area $c_t\lambda^2$, where c_t is constant.

A3. I/O gates have a minimum area $c_I\lambda^2$, where c_I is constant.

An additional assumption on I/O gates is required because of the different technology. Moreover, in general the relation $c_I \gg c_t$ holds.

A4. The number of wires which can be overlapped on any point in the layout is upper bounded by a constant n_f.

Technological advances make three-dimensional chip design approachable. As a consequence the number of layers (and hence the number of overlays) grows arbitrarily. It follows that assumption A4 makes sense in the present technology.

A5. The circuit area cannot be less than the sum of the areas dedicated to wires, transistors and I/O gates.

The previous assumptions are linked to problems for building circuits rather than for using them. One of the main problems arising from the use of circuits is the evaluation of the time needed for a signal to propagate from one transistor to another. Some studies [BPP81], [MC80], [Ul83], show the feasibility of the following assumption on time.

T1. Let $T(\mu)$ be the time necessary to transmit a signal between the two ends of a wire of length μ. We have $T(\mu) = O(1)$.

Assumption T1 (corresponding to a *synchronous* computing model) is the most used in the literature. The hypothesis that $T(\mu)$ is constant, i.e. independent of μ, is reasonable for small values of μ. If this is not the case, it is always possible to force circuits to follow T1 by means of synchronization tolls. Discussions and motivations for other assumptions can be found in [CM81].

P1. Every input data is read only once (*semellective* protocol).

This assumption takes into account data storage costs. If an algorithm needs to access more than a datum at a time, the datum must be stored. Some studies show that input duplication can provide better performance, both in terms of circuit area and computing time [KZ81].

P2. The I/O protocol is *unilocal*, i.e. each input or output data is available only in one location.

This protocol requires provisions for the links between an I/O gate and all circuit locations using the data passing through this gate. This takes into account the costs due to the fact that the same data can be used more than once during computations.

P3. Input and output data are available at given time units.
This protocol is called *time–determinate*.

P4. Input and output data are available at given locations.
This protocol is called *place–determinate*.

P5. The I/O protocol is word–local.

This axiom indicates that data are structured in words and all bits relative to the same word must be inserted in the circuit from the same I/O gate.

Some studies on I/O protocols in VLSI assume also that input/output occurs on the circuit boundary. In this case the circuit is called a *boundary chip*. When any location (even inside the chip) can be used for I/O, the circuit is called *general chip*. The latter assumption is reasonable if we think of the third dimension.

The VLSI computing model that we have described corresponds to the model that Ullman [Ul83] called the *grid model*. Geometrically, this model consists of a rectangular grid with square meshes. The computing elements can be placed on the intersection points, whereas wires can be placed along the border of the meshes. Criteria for element layout must respect the design rules introduced in Section 6.1. The grid model produces a rectangular circuit representation which is common in practice. All the circuits described in Section 6.4 have this characteristic.

6.3 Area–time complexity

One of the purposes of computational complexity is to specify (upper and lower) bounds on the cost of solving a problem. In general it is more difficult to obtain non–trivial lower bounds. This aspect is studied in this section, whereas in the following section computational complexity is applied to particular problems, for which we shall also describe circuits (hence upper bounds). In order to deal with the complexity of the model described in Section 6.2, it is necessary to consider its crucial resources. It is easy to realize that the crucial resources are circuit speed and size. The speed can be evaluated in terms of the computing time (T) of the algorithm implemented by the circuit. The circuit size is better expressed by the area (A) rather than the number of active components, because the space occupied by wires is often more than the space used by active components. An additional interesting measure is the period (P), that is, the minimum time interval between the solution of two instances of the same problem. Obviously, $T \geq P$, so that lower bounds on the period also apply to time.

Hereafter we shall consider a circuit as the layout of a computation graph obtained following the design rules of Section 6.1 and the axioms of Section 6.2.

6.3.1 Generalization

It is now appropriate to introduce the simple formalization that we assign to the terms "problem" or "computational problem" in the context of VLSI complexity.

A problem R is represented as a set of m Boolean functions of n arguments, i.e. R is viewed as a mapping:

$$R: (0, 1)^n \to (0, 1)^m$$

On the basis of the previous considerations, given a Boolean function to be computed, the problem arises of determining a VLSI circuit minimizing at the same time A, T and P, that is, using the minimum amount of all crucial resources. In general, such a problem does not have a solution. It is then necessary to introduce the cost function c, which depends on parameters A, T and P, i.e. $c = c(A, T, P)$. It is reasonable to suppose that such a function satisfies the following relations:

$$c(A', T', P') \leq c(A'', T'', P'') \text{ if } A' \leq A'', T' \leq T'', P' \leq P''$$

$$c(\alpha A, \tau T, \pi P) = g(\alpha, \tau, \pi) \, c(A, T, P)$$

where α, τ, π are positive constants, and g is an increasing function with respect to each argument. Under these hypotheses it is reasonable to choose a cost function such as:

$$c(A, T, P) = kA^r \, T^s \, P^q, \, k > 0, \, r, s, q \geq 0$$

Among these functions, the most used in the literature are AT and AT^2. Because of the nature of the cost functions best suited for VLSI computations, VLSI complexity is considered as *area–time* complexity.[2]

It is now possible to introduce some simple lower bounds, which will be useful in determining the complexity of problems considered below. These limitations derive either from general considerations on computations, or from properties of the model of the previous section.

1. A circuit with n memory cells requires an area $A \geq n$ and can store at most 2^n distinct states.
2. The time T required to compute a function for which at least one output depends on n inputs must satisfy $T = \Omega(\log n)$.

---

[2] A slightly different approach to VLSI complexity which is more useful for searching upper bounds (circuits) has been presented by Bilardi [Bi84]. Given a computational problem R, Bilardi associates to each circuit solving it (with input size n) in time T_0 and area A_0, a point with coordinates (T_0, A_0) in a plane which is called the *time-area plane*. The pair (T_0, A_0) represents the performance of the circuit. The set of all circuits solving R corresponds to a region of such a plane, which we shall call *feasible region*, the determination of which is the objective of VLSI complexity study.

The first characteristic of the permissible region is that if (T_0, A_0) is in it, the same applies also to all points (T_0, A) with $A > A_0$. In other words, if there is a circuit solving R in time T_0 and area A_0, then there are also circuits acting as the previous one, "wasting" additional area. This fact allows one to formulate the objective of VLSI complexity as follows: given a problem R, its VLSI complexity is described by the family of functions $A = <g_n(T)>_{n \in N}$, one for every n, where $<g_n(T)>_{n \in N} = \min\{A_0:$ a circuit exists solving R with performance $(T, A_0)\}$.

There are a minimum and a maximum time value ($T_{\min}(n)$ and $T_{\max}(n)$) defining the interval of interest. $T_{\min}(n)$ represents the time value under which it is not possible to have a circuit solving R. $T_{\max}(n)$ represents the time value after which $g_n(T)$ is constant, i.e. execution times greater than $T_{\max}(n)$ do not provide area reductions. Summarizing, given R we want to determine $g_n(T)$, such that $T \in [T_{\min}(n), T_{\max}(n)]$. In some cases it is possible to determine the order of magnitude of $g_n(T)$. In other cases the determination of upper and lower bounds must satisfy. The first situation arises when there are, for every T belonging to the interval of interest, circuits which are asymptotically optimum. The latter situation arises when there is a possible improvement in the order of magnitude for upper and/or lower bounds. It is easy to realize that $g_n(T)$ is an increasing function for n, and a decreasing function for T. This expresses the fact that a faster computation requires more hardware complexity.

3. The area and the time needed to compute a function with n I/O variables must simultaneously satisfy $AT \geq n$.
4. The area and the time needed to compute a function requiring (in sequential computing models) n_{op} arithmetic or logic operations must simultaneously satisfy $AT \geq n_{op}$.

It is now possible to describe one of the most widely used techniques for determining lower complexity bounds. This technique was first used by Thompson [Th79], [Th80] and subsequently extended by other authors [Bi84, BP86, Ya81]. The approach consists of partitioning the circuit in two parts, hence producing an *information flow* to be exchanged between the two parts of the circuit. Before analyzing the area–time complexity, let us analyze the concept of information flow (simply flow) which, as we will see, allows us to relate time and area in a natural way.

6.3.2 Flow

By information flow we mean the *motion* of data which is necessary to compute the function implemented by a circuit. There are two types of flow:

1. *the I/O flow* (data exchange between the circuit and the outside world);
2. *the internal flow* (data exchange inside a circuit).

The I/O flow can be:

1.1 *boundary flow* (in the case of a boundary chip, as in Figure 6.4)
1.2 *surface flow* (in the case of general chips, as in Figure 6.5).

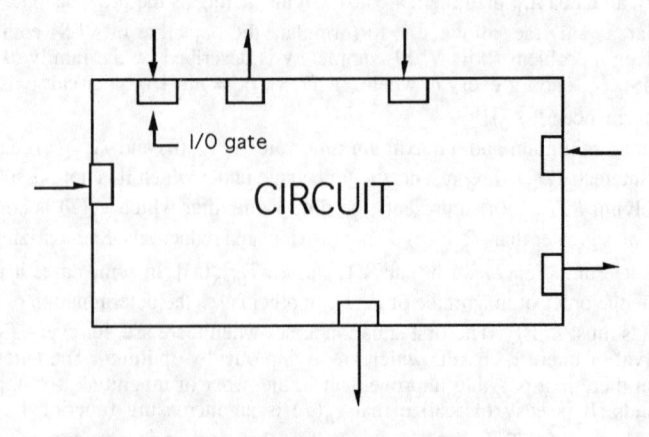

Figure 6.4 I/O flow in a boundary chip.

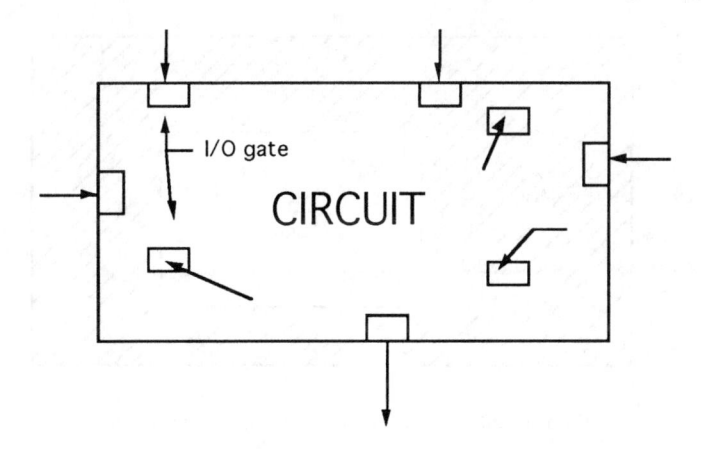

Figure 6.5 I/O flow in a general chip.

The I/O flow is also called primary flow, whereas the internal flow is called secondary flow. As we will see, there are problems where the primary flow is larger and vice versa.

In the case 1.1, the amount of information which can be exchanged with the outside world is proportional to the perimeter (rather than the area) of the circuit, so that the I/O process can be the bottleneck of the system. In fact, a circuit can contain a number of data proportional to the area but, at each instant, it cannot exchange with the outside more than an amount proportional to the perimeter. Since the case 1.2 is more general, it is appropriate to use it in determining lower bounds to complexity, as these limitations apply also to the more restrictive model.

Let us now consider a closed region inside a circuit, as shown in Figure 6.6(a). The resulting circuit appears to be partitioned in an internal region C_1 and an external region C_2. The boundary between the two regions is the area where information exchange (flow) takes place.

In this way we have *logically* divided the circuit into two systems communicating by means of wires crossing the boundary (Figure 6.6(b)). Recall that I/O operates through the circuit surface (using the general model for determining lower bounds), hence the set I of input variables is subdivided into I_1 and I_2, that is, in the subsets of variables corresponding to region C_1 and C_2 respectively. On the basis of assumption P1 (semellective protocol) I_1 and I_2 form a partition on I.

In the same way the set O of output variables is partitioned in C_1 and C_2. In general, the set $V = I \cup O$ of I/O variables is partitioned in $V_1 = I_1 \cup O_1$ and $V_2 = I_2 \cup C_2$. Moreover, the flow from C_1 to C_2 (from C_2 to C_1) is indicated with ϕ_{12} (ϕ_{21}). Figure 6.7 summarizes these concepts.

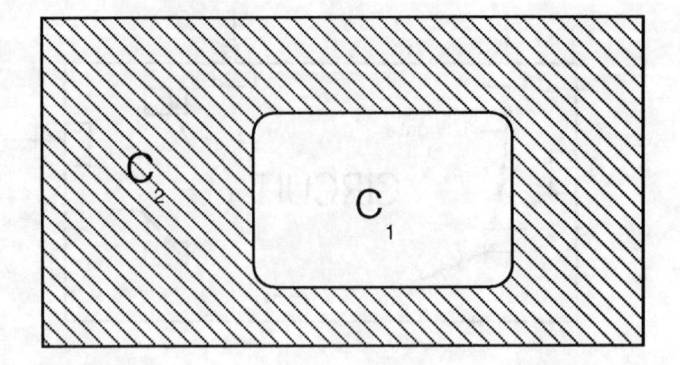

Figure 6.6(a) Partitioning a circuit in two regions.

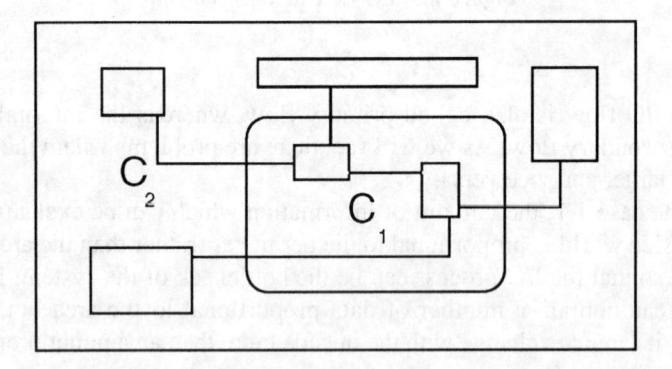

Figure 6.6(b) Logic partitioning of the system.

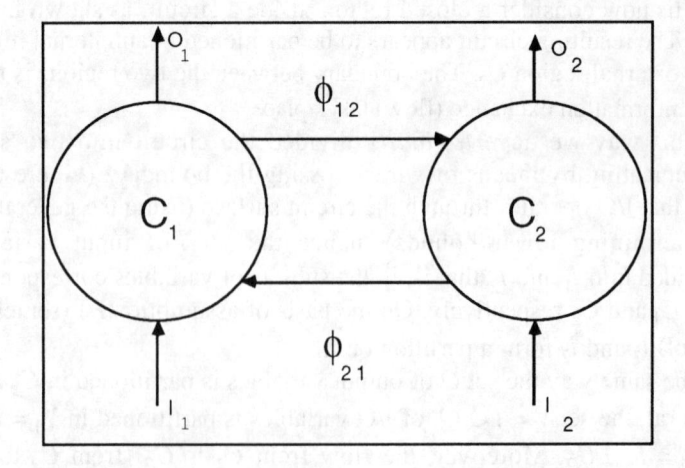

Figure 6.7 Partitioning and information flow.

Intuitively, the flow measures the number of information bits which must be transmitted between C_1 and C_2. The evaluation of the amount of data which must be exchanged between C_1 and C_2 will be the basis both for obtaining limitations to the time needed to perform the exchange, depending on the boundary perimeter and/or the number of wires crossing it, and for relating the flow to the measure AT^2.

Let (V_1, V_2) be a partition of the set V, which we call *assignment* because it induces a correspondence between I/O variables and C_1 and C_2. If we indicate with A_Q the algorithms solving problem Q, and with I_Q the instances of problem Q, then the flow necessary to solve Q is given by:

$$\min_{A_Q} \max_{I_Q} |\phi_{12}| + |\phi_{21}|$$

In other words, if the maximum of $|\phi_{12}| + |\phi_{21}|$ on all instances of Q is considered, then the minimum on all algorithms solving Q is taken.

In order to relate flow with area, it is useful to consider a circuit as a computation graph.

Consider the computation graph $G = (V, E)$ of an algorithm A_Q, where V and E are the set of vertices and arcs of G respectively. Assume further that A_Q solves Q in time T. Let (V_1, V_2) be a partition of V. It is then possible to determine the set of arcs connecting the two parts of the graph, hence defining the *separator width*, $\delta(V_1, V_2)$, as the number of arcs separating V_1 and V_2. In circuit terminology, $\delta(V_1, V_2)$ is a measure of the available *capacity* to exchange information between the two parts of the circuit. A partition (V_1, V_2) of the computation graph immediately induces an assignment of the I/O variables (V_1, V_2) to C_1 and C_2, simply by identifying V_1 with C_1 and V_2 with C_2. Note that the inverse is not true, because V cannot be arbitrarily partitioned; it is in fact possible that some elements of V use the same I/O gate. This problem is linked to I/O granularity, which can be formalized introducing a function $f: V \to N$, such that $f(v)$, for every $v \in V$, gives the number of variables treated from vertex (I/O gate) v during the computation. For the sake of simplicity, we now assume that $f(v) = 1$. The general case will be considered later.

For each computation graph of an algorithm solving a problem Q in time T, the following class of partitions is defined:

$$S_m = \{(V_1, V_2): |V_1| = m\}$$

where m is a positive integer. In a similar way the class of assignments to I/O variables, called H_m, is defined.

Let us now denote with $I_H(m)$ the *minimum flow* between C_1 and C_2, where the minimum is obtained by varying assignments in H_m (i.e. all possible assignments where one of the two subsets of variables has cardinality m); and similarly with $I_S(m)$ the *minimum separator* of graph G, where the minimum is obtained by varying partitions in S_m. Since at least I_H

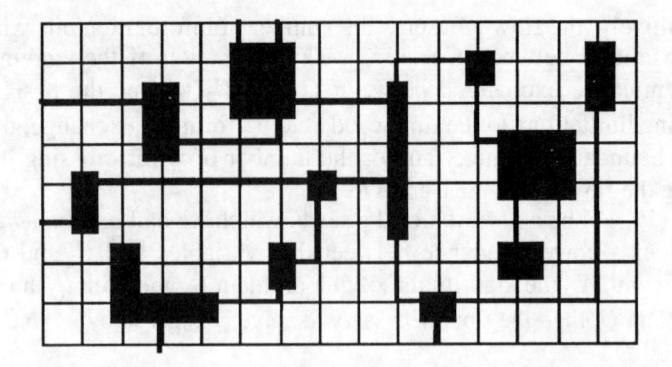

Figure 6.8 The grid model, introduced by Thompson in [Th80].

information bits must flow through the border between C_1 and C_2, we have that:

$$s(m) \geq I_H(m)/T \tag{1}$$

The inequality (1) is at the basis of an important lower bound to the measure AT^2 in terms of minimal flow. In order to obtain it, it is now necessary to relate the minimum separator with the circuit area. We shall do this by describing a technique called *tessellation* [BP86].

Assume that a VLSI circuit (or equivalently a layout for the computation graph) is a rectangle R on which a uniform grid is overlaid (Figure 6.8). This grid is called the *primary grid*. Its nodes correspond to computing elements (in a lower level description, to transistors) and its arcs correspond to interconnecting wires.

Let us now introduce a *secondary grid*. Its arcs intersect the arcs of the primary grid at half their length (Figure 6.9).

Let us finally impose on the two grids above, a square tessellation (Figure 6.10) with side l where:

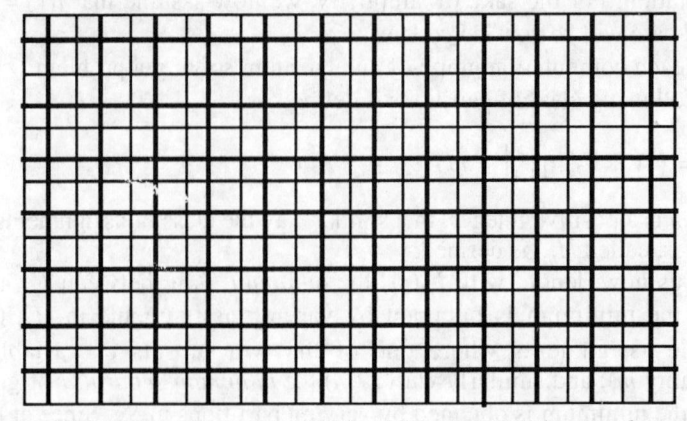

Figure 6.9 Primary and secondary grid (bold lines represent the secondary grid).

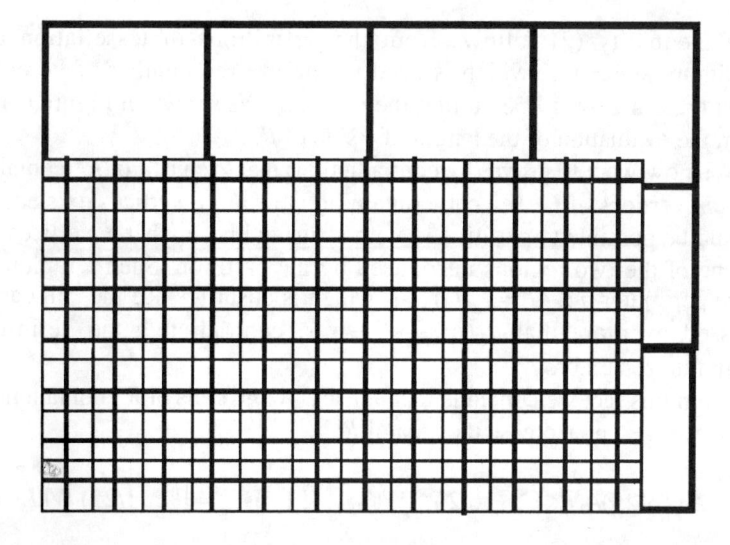

Figure 6.10 The tessellation grid, overlaid on a primary and secondary grid.

$$l = \lfloor (I_s(m) - 1)/4 \rfloor$$

Let now R^* be the smallest tessellation rectangle (formed by squares with side l) containing the circuit R. If $A(R)$ and $A(R^*)$ denote respectively the area of R and R^*, it is possible to prove that:

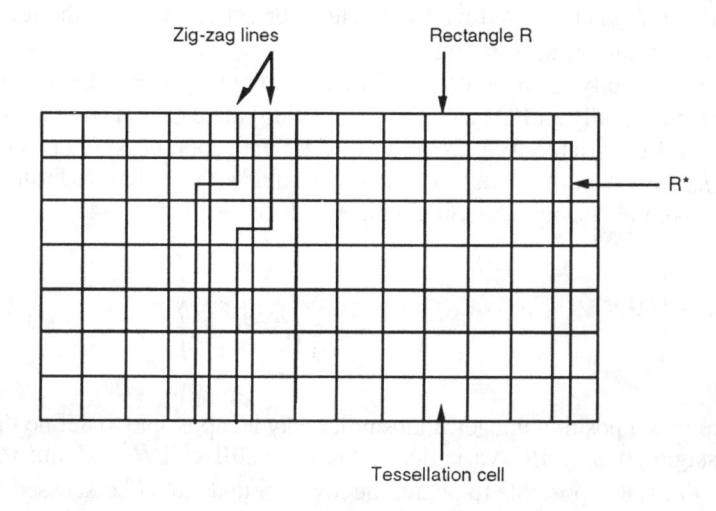

$$A(R^*) \leq (25/16) A(R) \tag{2}$$

Figure 6.11 Partitioning R with zig–zag lines.

The inequality (2) follows from the definitions of tessellation and of minimum separator (which is used to analyze the number of wires which must cross a zigzag line cutting the rectangle, as shown in Figure 6.11) and from the evaluation of the length of a side of R.

We now want to prove (by contradiction) that each cell of R^* contains less than m vertices of G. If a cell containing m or more vertices existed, then it should be possible to partition it, with a zigzag line, so that m vertices belong to one of the two regions introduced by the partition. Such a region would have perimeter $p \leq 4l = I_S(m) - 1$, which is absurd, since the cell cannot be crossed by more than $I_S(m) - 1$ wires, contradicting the definition of minimum separator.

From this fact we conclude that at least $\lceil |V|/m \rceil$ cells of R^* contain nodes of G, hence we can decrease the area of R:

$$A(R) \geq (16/25)A(R^*) \geq (16/25)\lceil |V|/m \rceil \lfloor (I_s(m)-1)/4 \rfloor^2 \geq (1/25)|V|I_s^2(m)/m$$

In conclusion, we have been able to diminish the circuit area in terms of the minimum separator. Combining this result with relation (1), we obtain:

$$AT^2 = \Omega(|V|I_H^2(m)/m) = \Omega(I_H^2(m)/m) \tag{3}$$

Many authors have used a technique known as *bisection* in order to determine lower bounds to AT^2 in terms of flow. Bisection leads to a lower bound which is a particular case of (3), i.e:

$$AT^2 = \Omega(I^2) \tag{4}$$

It is possible to obtain the inequality (4) starting from (3) and putting $m = n/2$, if $I_H(n/2)$ is indicated with I (the term bisection is due to the restriction $m = n/2$ on the induced partitions).

For simplicity we have used the assumption that $f(v) = 1$, i.e. that to each vertex (potentially an I/O gate in a general chip) there cannot correspond more than one I/O variable. We can now consider the general case. Let us indicate with f_{max} the maximum value of the function f. We must first redefine the set S_m to take into account the I/O granularity. It is:

$$S_m = \left\{ (V_1, V_2) : m - f_{max} + 1 \leq \sum_{v \in V_1} f(v) \leq m \right\}$$

where m is a positive integer. In a similar way it is possible to define the class of assignments to I/O variables, which we still call H_m. Using the new definitions, it is possible to obtain the lower bounds to AT^2 discussed above.

Summarizing, we have approached through tessellation the methods linking the square of the information flow to area and time. The particular case of bisection has been shown. It is an interesting possibility to evaluate the quantity $I_H^2(m)/m$ as a function of m and, for each problem, to search for the value of m maximizing $I_H^2(m)/m$, instead of limiting the analysis to values of m which are constant fractions of $|V|$, as in the case of bisection.

The limitations derived from flow analysis are useful in many cases. VLSI complexity for many problems is dominated by the request for data exchange. On the other side the flow does not exhaust all the possibilities. As we shall see, there are important computational problems where there are other factors dominating complexity. In other words, the techniques based on the notion of flow produce non–trivial lower bounds to the complexity of problems with a sufficiently high information flow. This happens, for example, when the links have a greater influence on the area than the computing elements. In other situations, different techniques seem more appropriate. This is the case with problems with a "strong request" for area (in terms of intermediate results to be stored) or when the area and the time needed to compute a function depend more on the number of operations needed to compute such a function rather than on the need to exchange information.

6.3.3 Saturation

Identifying a region in a circuit and logically isolating it is interesting not only because of the data flow which has to be exchanged with the rest of the circuit, but also to evaluate the amount of information which can be stored inside the region. This quantity is limited by the area of the region. Let us now consider a technique known as *saturation*, aiming at evaluating the influence of data storage on area–time complexity. For this technique we will still refer to the circuit subdivision of Figures 6.6 and 6.7.

The limitations on the amount of data which can be stored inside a region can influence information exchange. During the computation, one of the regions can be saturated, i.e. have no more space to store data. In this case, if there is a need for storing data, they should be sent to the other region for temporary storage. As soon as space is available, data are returned to the original region. Hence every bit involved in this process contributes twice to information exchange.

These considerations lead to a new definition of information flow in the case of memory limitations.

Suppose that C_1 and C_2 can store at most n_1 and n_2 bits respectively. The set S_m defined above is now $S_m(n_1, n_2)$, as well as H_m. As a consequence the definitions of *minimum flow* between C_1 and C_2, and of *minimum separator* of the graph G are extended. In this context the flow is denoted by $I(m; n_1, n_2)$. A flow without memory limitations is $I(m; \infty, \infty)$, whereas the flow when only C_1 has memory limitations is $I(m; n_1, \infty)$. Let us see how it is possible to combine the limitations on flow due to saturation with the

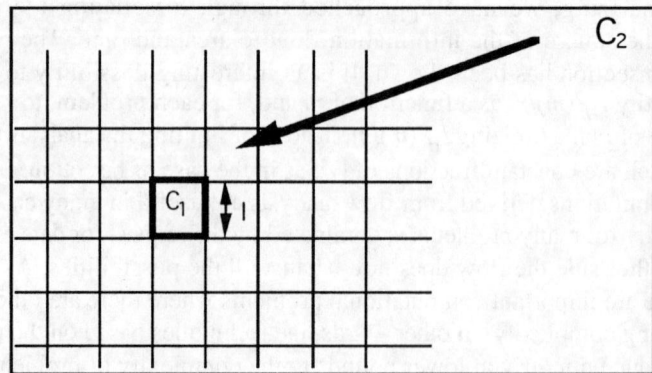

Figure 6.12 Circuit partitioning: region C_1 is a tessellation cell; region C_2 is the rest of the circuit.

tessellation technique. As shown in Figure 6.12, the tessellation cell (of side l) can be seen as a region C_1, and the rest of the circuit as region C_2. Obviously, the memory capacities of C_1 and C_2 are upper bounded by l^2 and $A - l^2$ respectively. If m variables are involved in I/O operations inside the cell, the information exchange along the border is $I(m; l^2, A - l^2) \geq I(m; l^2, \infty)$. Since the cell perimeter is $4l$, we have:

$$T \geq I(m; l^2, \infty)/4l \tag{6}$$

Since relation (6) applies for every m and l, the attempt is to determine m and l so that the lower bound is maximum. Let n be the number of I/O variables in a problem Q. Since N variables are inserted in (extracted from) the circuit with area A, then there exists at least a cell C_i of the tessellation managing m I/O variables, where $m \geq nl^2/A$. If it were possible to select a cell with $m = nl^2/A$, then it would be:

$$T \geq I(nl^2/A; l^2, \infty)/4l \tag{7}$$

A cell with these characteristics does not necessarily exist; however, it is possible to proceed as follows. Obviously in time T, at most T data can enter or exit a single I/O gate. Hence it is possible to divide cell C_i in the two regions, one of which manages exactly $nl^2/A + h$, $0 \leq h < T$, I/O variables. Moreover, the region perimeter and area are respectively at most $4l$ and l^2. It is then possible to prove that there exists an h such that:

$$T \geq I(nl^2/A+h; l^2, \infty)/4l, \ 0 \leq h < T \tag{8}$$

The fact that such a value of h exists leads to the following inequality:

$$T \geq \min_{0 \leq h < T} I(nl^2/A + h; l^2, \infty)/4l \tag{9}$$

In many cases the minimum value on the right–hand side of (9) corresponds to the case $h = 0$, which simplifies the evaluation of the flow. Sometimes the flow is of the order of the number of variables, i.e. $\alpha n l^2/A$, where α is a constant, so that $T \geq \alpha n l/4A$, hence we obtain:

$$AT \geq (\alpha/4)nl \qquad (10)$$

It is then possible to choose the value of l maximizing the right–hand side of (10). This value is often an increasing function of n, so that (10) is in many cases better than the limitation $AT = (n)$. This is a consequence of the fact that, for each problem with n I/O variables, $T \geq n/A$.

6.3.4 Functional dependence

Up to this point we have analyzed the information flow due both to computational reasons (the minimum flow, selected among all the computation graphs corresponding to algorithms solving the problem) and space reasons (data exchange to be temporarily stored). In both cases we have related flow with area and time: in the former case the square of the flow decreases the measure AT^2, whereas in the latter case the flow (when saturation is present) decreases the measure AT.

Let us now see a third technique, called *functional dependence of output on input*. It gives lower bounds to $AT/\log A$ [J80, Ba81, BP86]. Instead of focusing on the influence of communications on the area, we will now focus on the amount of memory required by the circuit.

This technique considers the dependencies between input and output with the consequent need for storing intermediate results. In general there are subsets of output variables depending on subsets of input variables. The computation of these output variables requires some time owing to the assumption that the computing elements (or logic gates) of the circuit have a fan–in limited by a constant (see Chapter 3). Hence the computing time cannot be less than the minimum depth of the computation graph, where the minimum is taken by varying all computation graphs corresponding to algorithms solving the problem. Output bits cannot exit the circuit before time has been spent reading the last input bit on which they depend. What happens is a "slow down" of the internal circuit flow due to the need to wait for data.

Let us denote the output (input) by a Boolean vector $\mathbf{o}(\mathbf{i})$ of order $r(s)$. The output is a function of the input, i.e. $\mathbf{o} = f(\mathbf{i})$. Let o_j be a component of $\mathbf{o}$ functionally dependent on i_j, i.e. changing i_j can cause a change of o_j.

Assume that:

- at time t_0, $i_1, i_2, ..., i_s$ are inserted in the circuit;
- $o_1, o_2, ..., o_r$ are dependent on each i_j, $j = 1, 2, ..., s$;
- $i_1, i_2, ..., i_s$ contain I information bits.

These assumptions have consequences on time and on circuit area:

- $T \geq \log s$ (for the fan–in arguments explained above, we have to wait for a time which is at least the minimum depth of a binary tree with s leaves);
- $A \geq I$ (during this time $\log s$, it is necessary to store the information contained in $i_1, i_2, ..., i_s$).

Combining these two results, we obtain a new lower bound to the measure AT, i.e:

$$AT \geq I \log s \qquad (11)$$

We can formalize this technique by means of the following approach.

Let U be a subset of input variables partitioned as: $U = \{U_1, U_2, ..., U_T\}$, where $U_j = \{$input variables inserted in the circuit at time $j\}$. Let $W = \{W_1, W_2, ..., W_T\}$ be a set of disjoint subsets of output variables, where W satisfies these conditions:

- each element of W_j has a functional dependence with a number of elements of U_j; indicate with $h(|U_j|)$ a lower bound to such a number;
- in order to compute the elements of set W_j it is necessary to have at least $I(|U_j|)$ information bits (contained in the input variables belonging to U_j).

Now we can repeat, for each pair (U_j, W_j), the considerations leading to (11) obtaining:

$$AT \geq \sum_{j=1}^{T} I(|U_j|) \log h(|U_j|) \qquad (12)$$

We had anticipated that the technique derived from the notion of functional dependence would have led to decreasing the measure $AT/\log A$. This is not apparent from (11) and (12), since the limitation is in terms of AT. However, many computational problems (e.g. integer addition [J80]) have the right–hand side of (12) which can be decreased with the product of an increasing function f of the size n of the input (and/or output), with $\log A$, hence obtaining $AT \geq f(n) \log A$, i.e:

$$AT/\log A \geq f(n) \qquad (13)$$

We have then finished the analysis of techniques for finding lower bounds to the VLSI complexity of computational problems. The methods illustrated up to this point are general ones, in the sense that they can be applied to any problem. For every problem it is necessary to select the most appropriate technique providing the best lower bound. Before considering the application of these techniques to specific problems (Section 6.4), we want to show the lower bounds to the complexity of two main classes of problems, the computation of transitive functions and the computation of bilinear forms.

6.3.5 Computing transitive functions

We now want to consider a result due to Vuillemin [Vu83a] concerning the area–time complexity for computing *transitive functions*. Let us first introduce some definitions.

Definition 6.1
A set S of permutations of the first n integers is called a *permutation group*, if the set S is a group.

Definition 6.2
A group G of permutations of the first n integers is called a *transitive group with degree* n, if, for each i, j, $1 \leq i, j \leq n$, there exists $g \in G$ such that $g(i) = j$.

Definition 6.3
A function f, $f(x_1, ..., x_n, y_1, ..., y_p) = (z_1, ..., z_n)$ is called *transitive with degree* n, if there exists a transitive group G of permutations on $\{1, 2, ..., n\}$ such that, for each $g \in G$, there exist binary values of $y_1, ..., y_p$ for which f is a permutation of $x_1, ..., x_n$ according to g, i.e:

$$f(x_1, ..., x_n, y_1, ..., y_p) = (x_{g(1)}, ..., x_{g(n)})$$

It is possible to prove that each circuit computing a transitive function with degree n must satisfy:

$$AT^2 \geq AP^2 = \Omega(n^2) \tag{14}$$

The inequality (14) can be proved by evaluating, by means of bisection, the flow necessary to compute a transitive function.

The class of transitive functions is very important. It includes the binary shift, the cyclic shift, the integers product and the convolution.

6.3.6 Computing bilinear forms

The solution of many computational problems is associated with computing *bilinear forms*. For example, the product of a matrix by a vector, the product of matrices, the product of integers and polynomials, the product of complex numbers, are all particular cases of bilinear forms.

In [CL87b] the bisection technique is applied to analyze the minimum information flow which is introduced to compute bilinear forms.

Let $A_1, A_2, ..., A_p$ be matrices of order $m \times n$, let **x** and **y** be vectors of order m and n respectively. Consider the set:

$$\{f_i: f_i = \mathbf{x}^T A_i \mathbf{y}, i = 1, 2, ..., p\} \tag{15}$$

The set defined in (15) is called the set of bilinear forms. A set of bilinear forms can be seen as a vector (in this case f, with components f_i). To every set of bilinear form is associated a *tensor* of the third order (informally, a matrix with three indices) **A**. Its element with position (h, k, l) is the element with position (h, k) on the matrix A_l introduced above. The matrices $A_1, A_2, ..., A_p$ are called 3-sections of tensor **A**. Similarly, it is possible to define the 1-sections and 2-sections of **A**, indicated respectively with B_j and C_i, $i = 1, 2, ..., n, j = 1, 2, ..., m$. The 1-sections and 2-sections of **A** are respectively $n \times p$ and $p \times m$ matrices.

It is known that the sequential complexity for computing (15) is equal to the rank of tensor **A**. As we will see, the VLSI complexity for computing bilinear forms can also be expressed in terms of properties of tensor **A**.

Let us now define two matrices $H(x)$ and $C(y)$ as follows:

$$H(\mathbf{x}) = \sum_{k=1}^{m} x_k B_k^T$$

$$C(\mathbf{y}) = \sum_{k=1}^{n} y_k C_k$$

The problem of computing (15) can be reformulated as the multiplication of matrix $H(\mathbf{x})$ by the vector **y**, or of matrix $C(\mathbf{y})$ by vector **x**. More precisely, we have:

$$\mathbf{f} = H(\mathbf{x})\mathbf{y} = C(\mathbf{y})\mathbf{x} \tag{16}$$

where the elements of the matrices are variables, whereas their structure is fixed.

Let us now consider a subdivision of the circuit in two regions C_1 and C_2, and the corresponding partition of I/O variables, $f = \{f_1, f_2\}$, $x = \{x_1, x_2\}$, $y = \{y_1, y_2\}$, such that $|f_1| = |f_2| = (1/2)p$, $|x_1| + |x_2| = m$, $|y_1| + |y_2| = n$.

If we also partition matrix $H(\mathbf{x})$ in terms of eight matrices $H_{rs}(x_t)$, $r, s, t = 1, 2$, then the computation of (15) in terms of (16) is:

$$f_1 = H_{11}(x_1)y_1 + H_{11}(x_2)y_1 + H_{12}(x_1)y_2 + H_{12}(x_2)y_2$$

$$f_2 = H_{21}(x_1)y_1 + H_{21}(x_2)y_1 + H_{22}(x_1)y_2 + H_{22}(x_2)y_2$$

It follows that the information which has to be exchanged through the partition is contained in the terms:

$$H_{11}(x_2)y_1, H_{12}(x_1)y_2, H_{12}(x_2)y_2, H_{21}(x_2)y_1, H_{22}(x_1)y_2 \text{ and } H_{21}(x_1)y_1$$

Consider as an example the term $H_{12}(x_2)y_2$. It is easy to realize that at least t information words must be transmitted from C_2 to C_1, where:

$$t = \max_{x_2} \text{rank}(H_{12}(x_2)) \tag{17}$$

Similarly, the transmission in the opposite sense (from C_1 to C_2) is expressed by the terms $C_{12}(y_1)$ and $C_{21}(y_2)$. Observing that $C_{12}(y_1)x_2 = H_{11}(x_2)y_1$ and that $C_{21}(y_2)x_1 = H_{22}(x_1)y_2$, and repeating the analysis leading to (17), it is possible to decrease the flow I as follows:

$$I \geq \max_{x_2 x_1} \text{rank}(H_{12}(x_2)) + \max \text{rank}(H_{21}(x_1)) +$$

$$+ \max_{y_2 y_1} \text{rank}(C_{12}(y_1)) + \max \text{rank}(C_{21}(y_2)) \tag{18}$$

6.4 VLSI complexity of some computational problems

In this section we will focus our attention on some computational problems chosen because of their importance and the differing characteristics they have with respect to VLSI complexity. We will first discuss (Sections 6.4.1 and 6.4.2) two arithmetic problems, integer sum and multiplication. For these problems the functional dependence of the outputs from the inputs and the flow are the aspects which, respectively, dominate complexity. Later (Sections 6.4.3 and 6.4.4) we will analyze the problem of the multiplication of a matrix by a vector, for which we will see that the primary flow (I/O flow) together with the choice of I/O protocol determine the area–time complexity. Finally, the multiplication of two matrices will be discussed, for which, on the other hand, the secondary flow determines complexity.[3]

---

[3] In determining upper bounds, in some instances, we will present a single circuit C of area A and time T, with the understanding that this indicates a set of circuits $\{C_n\}$, of area A_n and time T_n when the problem dimension n changes. In other cases, we will describe circuit

6.4.1 Integer addition

In this section we will analyze the area–time complexity of the computation of the sum of n digit integers represented in binary arithmetic. Let us first consider the problem of determining lower complexity bounds, following the work of Johnson [J80], together with the more general characterization seen in Section 6.3, in terms of the functional dependence between input and output.

Lower bounds

Let $a = a_n \ldots a_1$ and $b = b_n \ldots b_1$ be the binary representations of the numbers to be added and let $s = s_{n+1} s_n \ldots s_1$ be their sum. Let $A = A(n)$ be the area of the VLSI adder. Let $t_0, t_1, \ldots$ be a sequence of instants in time and let t_k be the first integer time value for which at least $3kA$ output bits have already been computed by the circuit. In the interval $[t_k, t_{k+1}]$ at least $3A$ outputs must be computed, and at most A of these can be computed at time t_k, since the number of bits which can be (simultaneously) stored in the circuit cannot exceed the area of the circuit itself. It follows that at least $2A$ outputs must be computed in the interval $(t_k, t_{k+1}]$. During such interval therefore the circuit outputs at least 2^{2A} possible different values.

Let now m be the number of bits read in the interval $(t_k, t_{k+1}]$. The outputs computed in the interval $(t_k, t_{k+1}]$ depend on the bits stored in the circuit at time t_k (which are at most A) and on the m bits read. One has then 2^{A+m} possibilities, from which it follows that $m \geq A$. The most significant of the new bits depends on all data used to compute the preceding outputs, and is therefore a function of at least A variables read after time t_k.

It is then:

$$t_{k+1} \geq t_k + \max\{\log A, 1\}$$

from which, letting $t_0 = 0$, one has:

$$t_k \geq k \max\{\log A, 1\}$$

Let now k_0 be the greatest integer for which $3k_0 A \leq n$. The following inequality holds:

$$n/3A - 1 \leq k_0 \leq n/3A$$

from which the bounds below follow:

$$AT \geq n, \qquad \text{for } A = 1$$

---

families, where this denotes a class C_T, made of sets of circuits $\{C_n\}$, of area $A_n(T_n)$ as the dimension n changes, and where T_n covers a given set of time values.

$AT/\log A \geq n$, for $A \geq 2$

This analysis led us then, except in the case when $A = 1$ (sequential addition), to a lower bound on the measure $AT/\log A$.

Upper bounds

Let us now see the design, proposed by Brent and Kung [BK82], of VLSI circuits optimal with respect to $AT/\log A$. As in the previous section, let $a = a_n \ldots a_1$ and $b = b_n \ldots b_1$ be the numbers to be added, and let $s = s_{n+1} s_n \ldots s_1$ be their sum. The usual method to compute the sum consists in applying the following relations:

$$c_0 = 0$$
$$c_i = (a_i \wedge b_i) \vee (a_i \wedge c_{i-1}) \vee (b_i \wedge c_{i-1}), i = 1, \ldots, n$$

$$s_i = a_i \oplus b_i \oplus c_{i-1}$$
$$s_{n+1} = c_n$$

where c_i is the carryover from position i.

A (similar) alternative scheme is the following:

$$c_0 = 0$$
$$c_i = g_i \vee (p_i \wedge c_{i-1}), i = 1, \ldots, n, \text{ where}$$
$$g_i = a_i \wedge b_i \quad \text{(carryover generation condition)}$$
$$p_i = a_i \oplus b_i \quad \text{(carryover propagation condition)}$$

Let us now define an operator op as follows:

$$(g,p) \; op \; (g',p') = (g \vee (p \wedge g'), p \wedge p')$$

for each Boolean variable p, g, p', g'.

Let:

$$(G_i, P_i) = \begin{cases} (g_i, p_i) & \text{if } i = 1 \\ (g_i, p_i) \; op \; (G_{i-1}, P_{i-1}) & \text{if } 2 \leq i \leq n \end{cases}$$

The relation $c_i = G_i$, $i = 1, 2, \ldots, n$ holds. Furthermore, it is easy to see that the operator op is associative. From this it follows that:

$$(G_i, P_i) = (g_i, p_i) \; op \; (g_{i-1}, p_{i-1}) \; op \; \ldots \; op \; (g_1, p_1)$$

234 VLSI algorithms and architectures

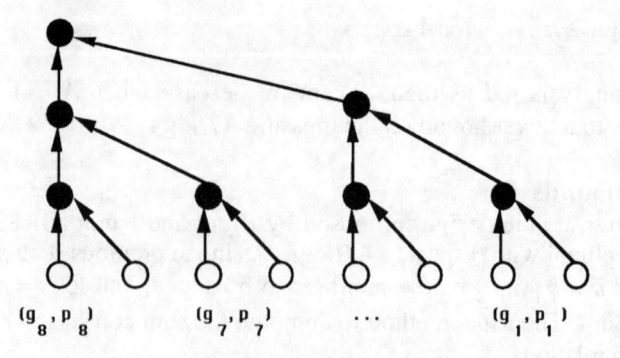

Figure 6.13 Tree scheme to compute (G_8, P_8).

Consider now the problem of computing (G_n, P_n). From Figure 6.13, illustrating the case $n = 8$, it is apparent that for such a subproblem we obtain the upper bounds:

$$A = O(n\log n), T = O(\log n)$$

Figure 6.14 shows the circuit for the computation of all carryovers, in the case $n = 8$. It is clear that the bounds $A = O(n\log n)$, $T = O(\log n)$ still hold.

Now let k be the number of bits of each operand, simultaneously accepted by an adder, with $k < n$. Note that the situation in Figure 6.14 refers to the case $k = n$. Let us partition an n-bit integer into n/k segments, each one containing k consecutive bits. Starting from the k least significant bits, it is easy to see how the computation of the first k carryovers can be done using the layout in Figure 6.14, and how the carryover computation for successive segments can be done in a pipeline. The results thus produced are not the final carryovers however. The partial results corresponding to the ith segment must be modified by applying $(G_{(i-1)}, P_{(i-1)k})$ through the operator op.

To execute this it is possible to insert a further tree structure (Figure 6.15) in the upper half of the layout of Figure 6.14, that is, starting from the fourth tree level. The following performances are obtained:

$$A = O(k\log k), T = O(n/k + \log k)$$

that is:

$$AT/\log A = O(n), \text{ for } k \in [\Theta(1), n/\log n], \text{ i.e. } T \in [\Theta(\log n, n)]$$

The design method we have just described allows one to reach the lower bound for every degree of parallelism.

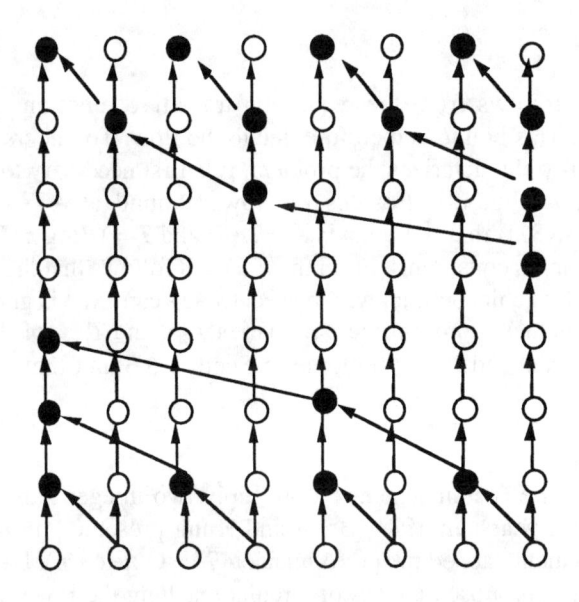

VLSI complexity of some computational problems **235**

Figure 6.14 Scheme to compute carryovers for $n = 8$.

6.4.2 Integer multiplication

In this section we analyze the VLSI complexity of integer multiplication. To be exact, we will demonstrate a technique, based on the *divide et impera* principle, that applies directly to the product of two polynomials, and that can be immediately adapted to integer multiplication.

Lower bounds
For the problem of computing the product of n–bit integers, the lower bound:

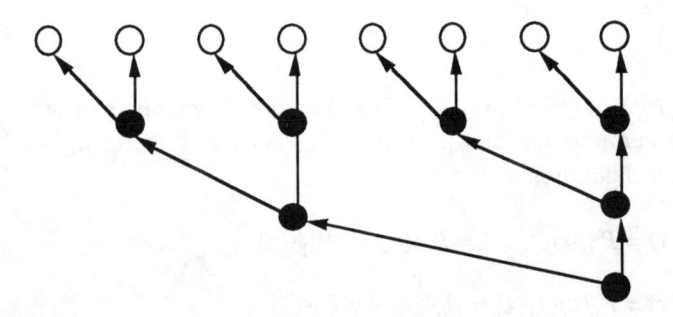

Figure 6.15 Tree structure, to be inserted in the upper half of the layout of Figure 6.14, allowing the computation of all carryovers.

$AT^2 = \Omega(n^2)$

holds. This follows from the results on transitive functions presented in Section 6.3. This bound is therefore due to the flow. To understand whether the flow really characterizes the problem, it is first necessary to evaluate the existing upper bounds. Besides the lower bound at AT^2, one can see ([AA80], [BK81]) that the bounds $A = \Omega(n)$ and $T = \Omega(\log n)$ hold, the last one due to fan–in constraints. From these facts it follows that the time interval within which circuits optimal with respect to AT^2 can exist is given by ($\log n$, $\sqrt{n}$). These values represent the boundaries T_{min} and T_{max} of the time–area plane we introduced in the footnote in Section 6.3 and that will limit our attention.

Upper bounds

The search for efficient schemes to multiply two integer numbers in VLSI lasted several years. In 1981, Brent and Kung presented an n-bit integer multiplier which reached the performance $AT^2 = O(n^2 \log^3 n)$. Later Preparata and Vuillemin proposed a class of circuits reaching the lower bound $AT^2 = \Omega(n^2)$ for every value of time between $\log^2 n$ and $\sqrt{n}$. An optimal circuit, with computation time less than $\log^2 n$, had not been found yet, although parallel algorithms existed taking logarithmic time. After that, circuits were developed with a logarithmic computing time, although not optimal with respect to AT^2. In 1983, Mehlhorn and Preparata [MP83] finally presented a circuit optimal for every time value. Their circuit uses the fast Fourier transform on a Fermat ring. The existence of optimal circuits ensures that the flow dominates the complexity of the multiplication problem. In this section we will restrict ourselves to presenting a circuit with less than optimal performance, but easy to implement. As anticipated, the problem faced is that of the product of two polynomials with a *divide et impera* approach.

Let P and Q be two polynomials of order less than or equal to $2n$. Let C be the polynomial product of P and Q. One can write:

$$P(x) = P_0(x) + x^n P_1(x)$$
$$Q(x) = Q_0(x) + x^n Q_1(x)$$
$$C(x) = C_0(x) + x^n C_1(x) + x^{2n} C_2(x) + x^{3n} C_3(x)$$

where $P_0, P_1, Q_0, Q_1, C_0, C_1, C_2$ and C_3 are polynomials of order less than n. The computation of C_i, $i = 0, 1, 2, 3$, can be done using the following recursive algorithm:

$$P^1(x) = P_0(x) Q_0(x) = P_0^1(x) + x^n P_1^1(x)$$

$$P^2(x) = P_0(x) Q_1(x) = P_0^2(x) + x^n P_1^2(x)$$

$$P^3(x) = P_1(x) Q_0(x) = P_0^3(x) + x^n P_1^3(x)$$

$$P^4(x) = P_1(x)Q_1(x) = P_0^4(x) + x^n P_1^4(x)$$

from which:

$$C_0(x) = P_0^1(x)$$

$$C_1(x) = P_1^1(x) + P_0^2(x) + P_0^3(x)$$

$$C_2(x) = P_1^2(x) + P_1^3(x) + P_0^4(x)$$

$$C_3(x) = P_1^4(x)$$

Following [Ba83] let us now show how the recursive algorithm above can be translated into a circuit design. Let M_n be the circuit executing the multiplication of two polynomials of order less than n. Let p_i, q_i and c_i be the coefficients of P, Q and C respectively. These coefficients are encoded by k bits. The circuit M_n has $2n$ input cables, one for every coefficient, and therefore the bits of each coefficient are read in series. The output cables are $2n - 1$, corresponding to the coefficients $c_0, c_1, ..., c_{2n-2}$ of C. The basis of the recursion is a circuit M_1 executing the serial product of two k-bit integers, p and q, computing the k bits of $c = ab \bmod 2^k$ (see Figure 6.16(a)).

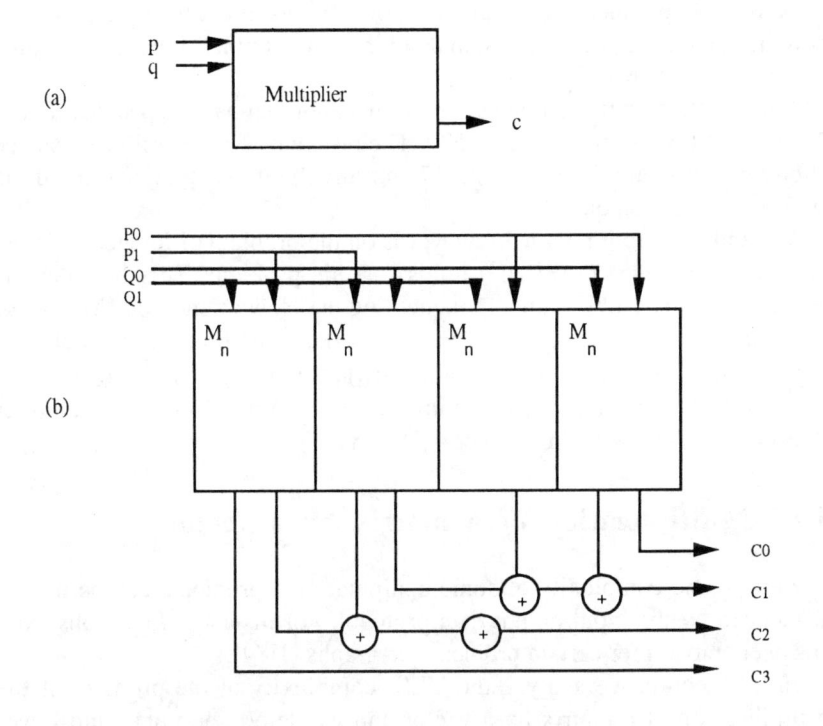

Figure 6.16 Multiplication circuit: (a) k-bit integer multiplier; (b) recursive circuit.

The formulas shown above can now be used to build a circuit M_{2n}, starting from four M_n circuits and four adders, as shown in Figure 6.16(b). Now let H_{2n} and L_{2n} be the vertical and horizontal sizes respectively of M_{2n}. The relations:

$$H_{2n} = L_n + 4n$$
$$L_{2n} = 4H_n$$
$$T_{2n} = T_n + kt$$
$$P_{2n} = \max\{P_n, kt\}$$

hold, where kt is the time needed to execute serial addition. Solving the recursive equations one has:

$$AT^2 = O(n^2 \log nk)$$
$$AP^2 = O(n^2 \min\{k, \log^2 n\})$$

This result does not reach the lower bound we have seen in the previous section.

A recursive circuit optimal with respect to the measure AP^2 and less than optimal (by a logarithmic factor) with respect to AT^2 has been obtained (see [LV83]) using the Karatsuba algorithm [KO62]. This algorithm enables one to express the product of two degree n polynomials in terms of three, instead of four, degree $n/2$ polynomials, in combination with the Muller and Atrubin multiplier [At65, Mu63, MP75].

It is well known that all integer multipliers optimal with respect to AT^2 use the discrete Fourier transform (DFT). The problem of the possible existence of circuits optimal with respect to AT^2 not involving the computation of the DFT is still an open question.

We end this section with a few words on the problem of integer division. The lower bound $AT^2 = \Omega(n^2)$ holds for this problem. This boundary is reached by a class of circuits implementing the Newton method [Me84] for time values between $\log^2 n$ and n, and further (for time values between $\log^{1+\varepsilon} n$ and n, for any $\varepsilon > 0$) by a somewhat complicated circuit [MP85] implementing, with suitable changes, the Beame, Cook and Hoover algorithm we have seen in Chapter 4 [BCH84].

6.4.3 Multiplication of a matrix by a vector

The area–time complexity of some computational problems can be deeply influenced by the input/output data protocol. For instance, Ja' Ja' observed this phenomenon for certain problems on graphs [JJ84].

In this section we analyze the VLSI complexity of the problem of the multiplication of a matrix by a vector and we determine upper and lower bounds which will be dependent on the I/O conventions.

Following the approach presented in [CLR90], we will consider different possibilities regarding the structure of I/O devices:

G. the I/O ports can lie on any chip location;
B. all I/O ports lie on the chip boundary;
P. it is allowed to introduce data in pipeline through the I/O ports;
N. it is not allowed to introduce data in pipeline.

These alternatives enable one to introduce four different I/O protocols, which will be denoted below by GP, GN, BP, BN. Now let $A = (a_{ij})$ be a real $n \times n$ matrix and let $x = (x_j)$ be an order n vector. The problem we must analyze is computing the product $Ax = y$. We have two possibilities:

1. the matrix A resides in the circuit and the vector **x** is the input;
2. the matrix A and the vector **x** both constitute the input.

The first case is typical of situations when the elements of A are constant, and are therefore inserted in the circuit in a pre–processing phase, or are even hardwired in the circuit. This is for instance the case of the Fourier matrix, the elements of which have a fixed value, once the dimension is fixed (in other words, the value of the elements depends only on matrix size).

Lower bounds
Let us now see a few lower bounds, both of a general and of a specific (of validity limited to certain models) type. We will then go on to analyze design techniques enabling one to find optimal architectures. It is easy to see that the sequential complexity of a problem represents a lower bound with respect to the AT measure (see Section 6.3). n^2 arithmetic operations are needed to compute the product of an $n \times n$ matrix by an n–vector. From this it follows that $AT \geq n^2$, for any chosen I/O protocol. (The same bound could have been trivially obtained by observing that $AT \geq$ {number of inputs}, in the case when the matrix elements are input variables.) Furthermore, using the technique discussed in Section 6.3 or based on known analyses, one can easily see that the information flow for the matrix–vector product is greater than or equal to n. Thus, we have the following lower bounds:

1. Resident matrix
   $A \geq n^2$, independent of I/O protocols.
2. Non–resident matrix
   $AT \geq n^2$, in the GP model,
   $AT^2 \geq n^3$, in the BP model,
   $A \geq n^2$, in the GN model,
   $AT \geq n^3$, in the BN model.

The implementations we will show in the following section refer to case 2.

Upper bounds

Before giving a detailed description of the circuits solving the problem of the multiplication of a matrix by a vector, we will introduce a class of architectures widely used in executing VLSI matrix operations. These are *systolic* arrays, introduced for the first time by Kung and Leiserson in [KL80]. The VLSI technology made clear that simple interconnections lead to low-cost implementations, with high integration density and high performance. Systolic systems have these positive characteristics, since they simultaneously include the pipeline, parallelism and interconnection regularity concepts. A systolic system consists of a network linked by dedicated and regular interconnections. Each processor rhythmically computes and sends data through the system. Systolic systems represent a realistic computational model, as observed for instance in [Ku82], since the regular processor "pulse" leads to a regular data flow throughout the system. Every process is able to transmit data and to execute simultaneously a given number of simple (arithmetic or logic) operations. When solving problems of numerical linear algebra the most widely used systolic structures are the linear array, which is above all well suited to execute matrix-vector products as well as convolutions, and the mesh, which can be efficiently used to multiply two matrices. The basic operation executed by a processor of the systolic array to compute the matrix-vector product (and, as we shall see, to compute the product of two matrices) is summing to its own contents the product of the data it receives as input from the adjacent processors. Let us assume, following Kung and Leiserson, that every processor has three registers, each of which has two connections with the outside world, one to receive input data and one to send output results. Figures 6.17(a) and 6.17(b) show two examples of structures (square and hexagonal) which can be used as a scheme for the single processor layout. The type (a) structure is used to compute the matrix-vector product and to solve triangular systems, whereas the type (b) structure is used to compute the product between matrices and the LU decomposition [MC80]. The processing time unit of the system is usually

(a) Square processor
(4 connections)

(b) Hexagonal processor
(6 connections)

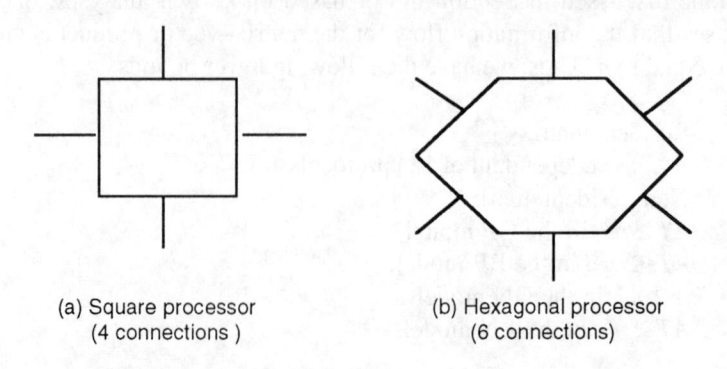

Figure 6.17 (a) Square processor; (b) hexagonal processor.

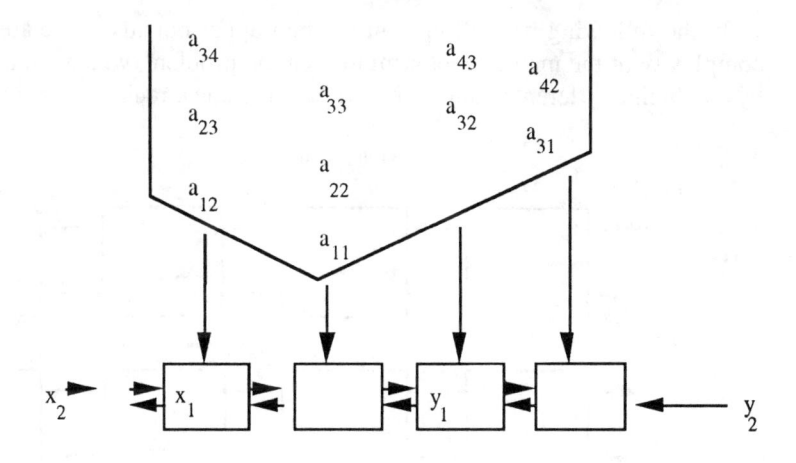

VLSI complexity of some computational problems 241

Figure 6.18 Systolic scheme to compute the product of a band matrix by a vector.

given in terms of the processing time of the single processor. Within each unit time interval, the processor executes a shift of the data present at its input connections and transfers them to its internal registers. Subsequently, it executes the operation and sends on the output lines both the data received as input and the computed result.

Let us now consider in detail the systolic solution to the problem of the computation of the matrix–vector product. The components of the vector y, which is the product of an $n \times n$ matrix A by an n–vector x, can be computed by means of the following recursive relations:

$$y^{(1)} = 0$$

$$y^{(k+1)} = y^{(k)} + a_{ik}x_k, \ k = 1, 2, ..., n; \ i = 1, 2, ..., n$$

$$y = y^{(n+1)}$$

Assume A is a band matrix with a bandwidth $b = p + q - 1$. In this case the above recursions can be computed by means of a linear systolic array made of b processors, executing the computation of y in pipeline. Figure 6.18 shows the operations executed by such an array when $p = 2$ and $q = 3$ (that is, $a_{ij} = 0$ for $i > j + 2$ and $j > i + 1$). The system consists of four processors, independent of matrix size. Figure 6.19 shows a few computing steps executed by the system.

242   VLSI algorithms and architectures

In the following we will present optimal upper bounds to the area–time complexity of the matrix–vector multiplication problem, which are obtained by combining different architectures within the same circuit.[4]

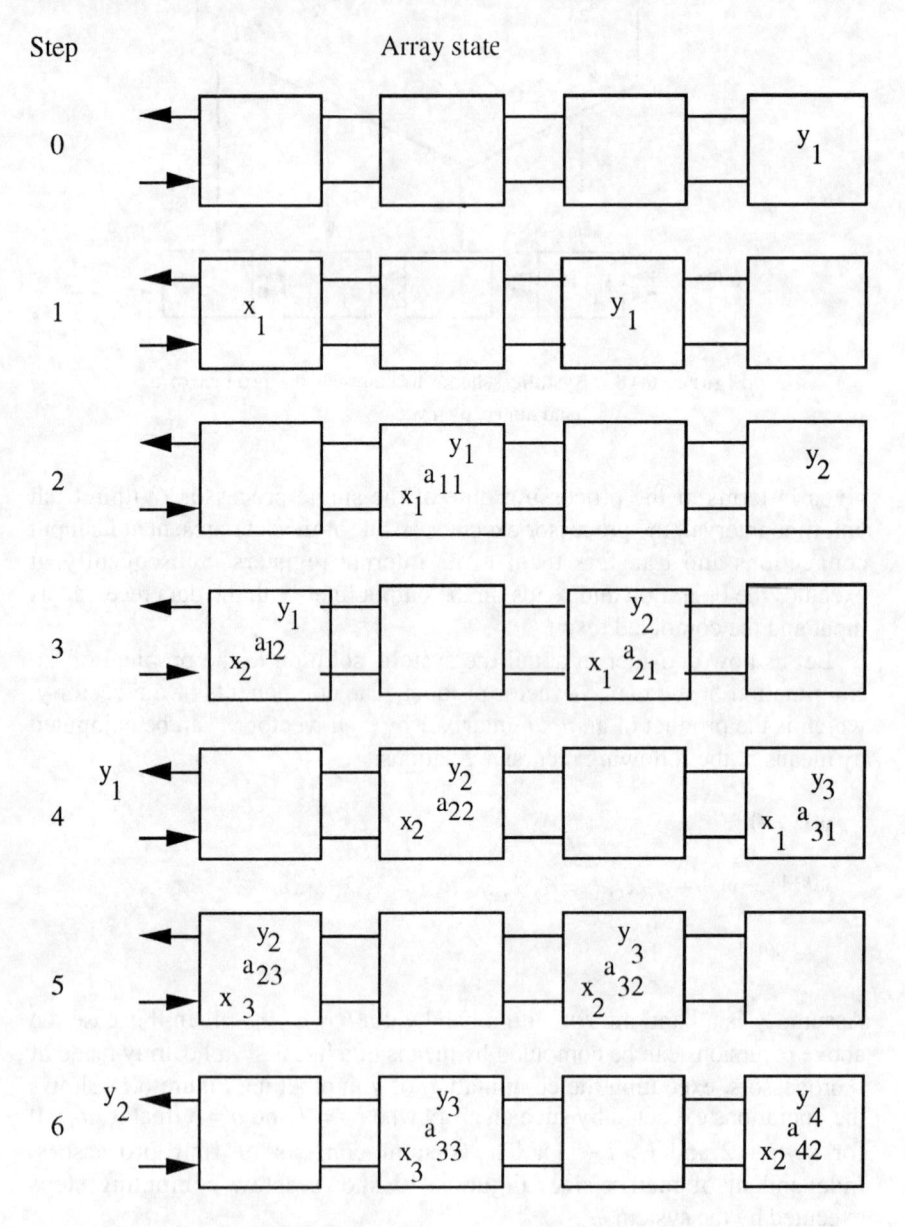

Step   Array state

0 y_1

1   x_1 y_1

2   y_1 y_2
                   a_{11}
                   x_1

3   y_1 y_2
     a_{12} a_{21}
     x_2 x_1

4 y_1 y_2 y_3
                         a_{22} a_{31}
                         x_2 x_1

5   y_2 y_3
         a_{23} a_{32}
         x_3 x_2

6 y_2 y_3 y_4
                         a_{33} a_{42}
                         x_3 x_2

Figure 6.19 The first computation steps, in the case shown in Figure 6.18.

---

[4]The technique used is similar to the one introduced in Section 4.1.5. If it is possible to describe recursively both the problem to be solved and the corresponding circuit, then a suitable minimum size is fixed below which different circuits are used.

VLSI complexity of some computational problems 243

The first instance we will consider is that relative to the "basic module" shown in Figure 6.20, consisting of an arithmetic processor and a finite memory. Such a module executes the trivial multiplication algorithm with the following performance.

1. Resident matrix
   $A = O(n^2)$, $T = O(n^2)$, independent of the I/O protocol.
2. Non–resident matrix
   $A = O(n^2)$, $T = O(n)$ in the GP and BP models,
   $A = O(n^2)$, $T = O(n^2)$ in the GN model.

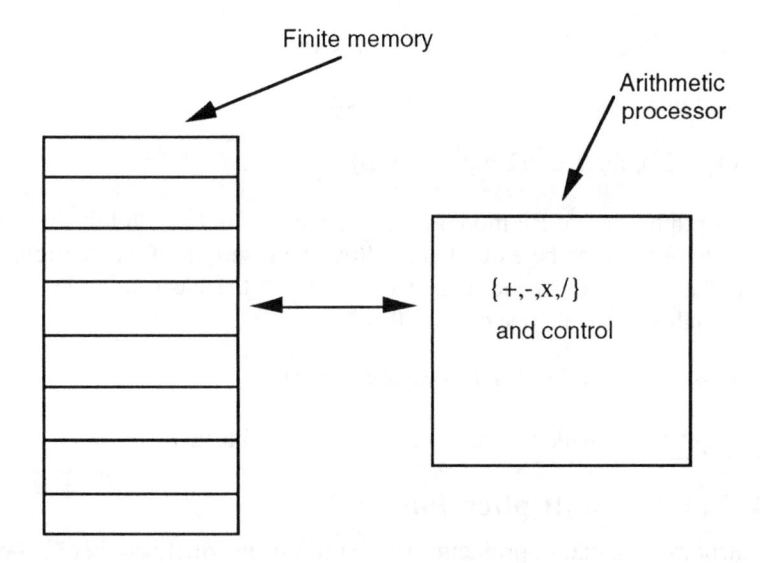

Figure 6.20 Basic module.

Clearly, the performance obtained is not optimal. The basic module however can be used as an elementary processor to execute order k ($k \ll n$) matrix–vector products in a mesh of trees computing the order n matrix–vector product. More precisely, one can see an $n \times n$ matrix as an $(n/k) \times (n/k)$ order k block matrix, execute the operations on the blocks in the elementary processors and then reconstruct the result by means of a data (k–vectors) flow through the mesh of tree connections. Usually, the matrix–vector product is computed in a mesh of trees the nodes of which execute scalar operations ($k = 1$ case). In this context, on the other hand, the mesh of tree nodes execute operations on $k \times k$ matrices. It is possible to prove that the performance of the mesh of trees with elementary processors is given by:

$A = O(n^2 A(k)/k^2 + n^2/k^2 \log^2(n/k))$
$T = O(\log(n/k) + k^2)$

In the GP and BP models, for k between $\sqrt{\log n}$ and $\log n$, one has:

$$A = O(n^2/T\log^2 n/T),\ T \in \{\Omega(\log n),\ O(\log^2 n)\}$$

which is optimal, provided logarithmic factors are neglected.

Another simple structure which can be used as a basic module to be inserted in a more complex circuit is the linear array. This has components which can be, as discussed above, simple processing elements capable of executing products between scalar quantities Such a structure is, among other things, optimal in the GP and BP models. The linear array can be used to execute order k matrix–vector products in a mesh of trees, in a way similar to the elementary processor. In the GP model, one has:

$$A = O(n^2/T^2\log^2 n/T),\ \text{for}\ T \in (\Omega(\log n),\ O(\log^2 n))$$

which is optimal, neglecting logarithmic factors, and

$$A = O(n^2/T^2),\ \text{for}\ T \in (\Omega(\log^2 n),\ O(n)),$$

which is optimal. In the BP model case, the mesh of trees with linear arrays as basic elements must be extended to allow the insertion of connections to transmit the matrix from the circuit boundary to the basic modules. In a somewhat laborious way, one can see that:

$$A = O(n^3/T^2\log^2(n/T)),\ \text{for}\ T \in (\Omega(\log n),\ O(n)),$$

which is optimal, provided logarithmic factors are neglected.

6.4.4 Matrix multiplication

The matrix multiplication problem in VLSI has been widely studied. Lower bounds with respect to AT^2 have been determined in [Sa79, Sa81]. [KL80] and [PV80] respectively presented a systolic architecture reaching the optimum with respect to AT^2, and a family of VLSI architectures reaching the lower bound for any processing time value between $\log n$ and n.

Lower bounds

Savage showed that, by applying techniques based on information, one obtains the following bound for the multiplication of $n \times n$ matrices: $AT^2 = \Omega(n^4)$. This bound is reached by a family of architectures which will be presented later. Matrix product is therefore a typical example of a problem where the area–time complexity is determined by the information flow, in particular by the secondary flow.

Upper bounds

This section is organized as follows. We first describe a systolic machine where each processor has a hexagonal structure (see Section 6.4.3) capable

VLSI complexity of some computational problems   245

> $A = (a_{ij})$, $a_{ij} = 0$ for $i > j+2$ and for $j > i+1$;
> $B = (b_{ij})$, $b_{ij} = 0$ for $i > j+1$ and for $j > i+2$;
> $C = AB$,
> $C = (c_{ij})$, $c_{ij} = \Sigma_k a_{ik} b_{kj}$.
> $c_{ij} = 0$ for $i > j+3$ and for $j > i+3$.

Figure 6.21 Multiplication of matrices with bandwidth 4.

of computing the product of two $n \times n$ matrices in $O(n)$ time and $O(n^2)$ area. We will then present a VLSI design by Preparata and Vuillemin reaching the lower bound for every degree of parallelism.

Systolic architecture   Let $A = (a_{ij})$, $B = (b_{ij})$, $C = (c_{ij})$ be three $n \times n$ matrices. The product $AB = C$ can be computed using the following recursions:

$$c_{ij}^{(0)} = 0$$
$$c_{ij}^{(k)} = c_{ij}^{(k-1)} + a_{ik} b_{kj}, \quad k = 1, 2, ..., n$$
$$c_{ij}^{(n)} = c_{ij}$$

for $i, j = 1, 2, ..., n$.

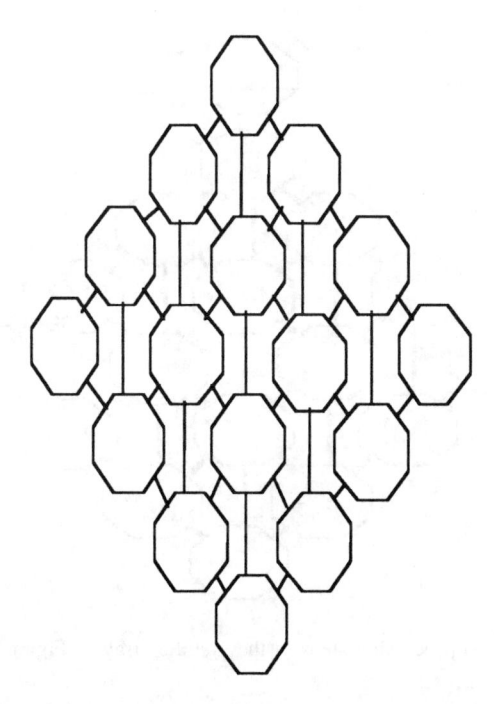

Figure 6.22 Systolic machine relative to the problem shown in Figure 6.21.

246 VLSI algorithms and architectures

Let us consider, for the sake of simplicity, the case when A and B are band matrices, with strip width b_1 and b_2 respectively. In particular, let us analyze the case shown in Figure 6.21, where the bandwidth is 4, both for matrix A and for matrix B.

In this case, the systolic machine consists of 16 processors, as shown in Figure 6.22. The first processing steps, clarifying the pipeline processing strategy being used, are shown in Figure 6.23.

The interesting aspect of this kind of implementation, besides arguments relevant to architectural simplicity, rests on the fact that the area of the system increases linearly with the band product of the matrices to be multiplied, independent of matrix size.

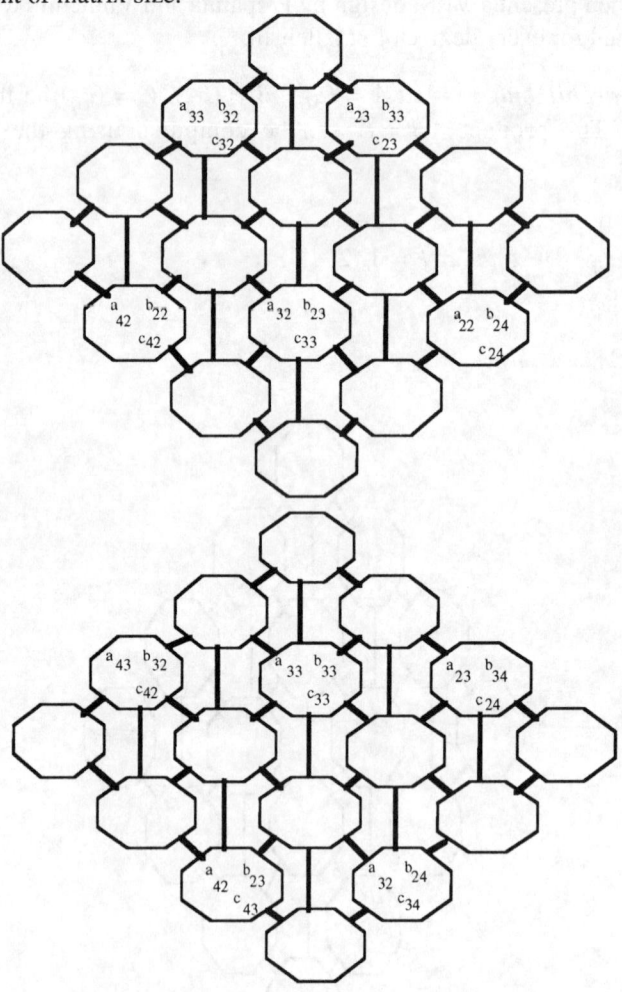

Figure 6.23 Two processing steps in the systolic array of Figure 6.22: data flow is evidenced.

Let us now show a design strategy, presented in [PV80], which enables one to obtain circuits reaching the lower bound for any value of T between $\log n$ and n.

Preparata–Vuillemin circuits. Let A and B be two $n \times n$ matrices. Consider, for A, the block decomposition:

$$A = \begin{pmatrix} A_{11} & A_{12} & \cdots & A_{1r} \\ A_{21} & A_{22} & \cdots & A_{2r} \\ \vdots & \vdots & & \vdots \\ A_{r1} & A_{r2} & \cdots & A_{rr} \end{pmatrix}$$

where A_{ij}, $i, j = 1, 2, \ldots, r$, are n/r matrices. A similar decomposition can be performed on matrix B.

Further, let $C_j = [A_{1j} \ldots A_{rj}]^T [B_{j1} \ldots B_{jr}]$, $j = 1, \ldots, r$. It is:

$$C = AB = \sum_{j=1}^{r} C_j$$

Consider now a mesh of processors. Using such a structure, the product between matrices can be computed by means of a pipeline scheme, as shown in Figure 6.24.

At each step, one mesh diagonal is activated at a time, and processing proceeds from the upper left corner to the lower right one. Therefore the computation can be entirely executed in pipeline, without the need to wait. Note that each mesh element must execute an operation of the type $X = X + YZ$, where X, Y and Z are $n/r \times n/r$ matrices.

The structure of a mesh element is shown in Figure 6.25, whereas Figure 6.26 presents the detailed circuit executing the product between n/r matrices.

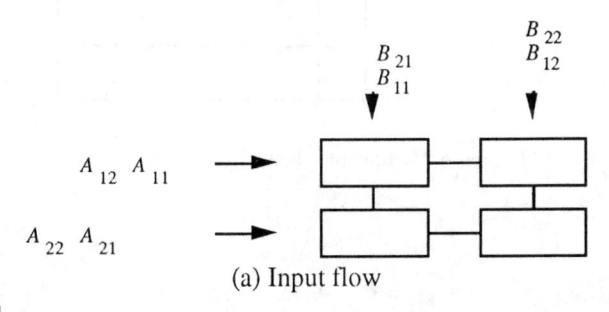

(a) Input flow

Figure 6.24

248 VLSI algorithms and architectures

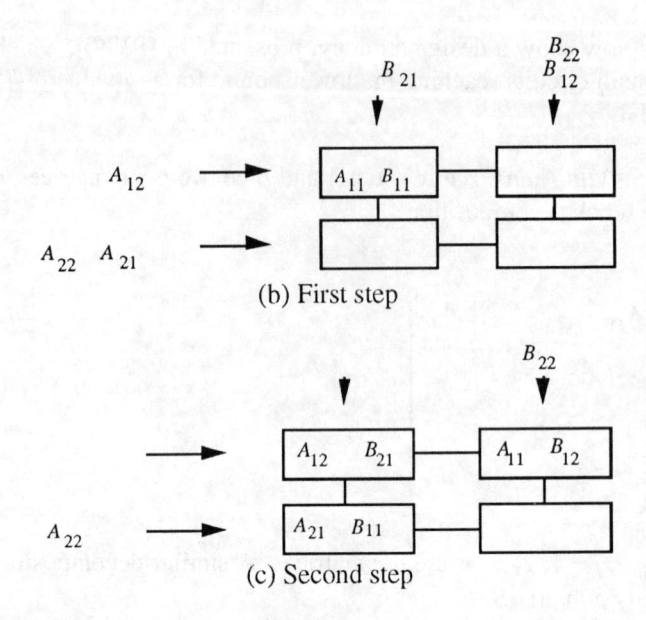

(b) First step

(c) Second step

Figure 6.24 Input and first processing steps in the pipeline scheme, when $r = 2$.

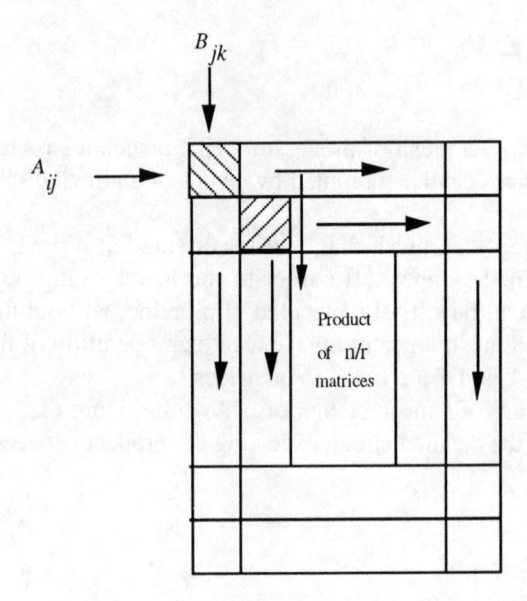

Figure 6.25 Layout scheme of a mesh element.

Each mesh element has $O(n^4/r^4)$ area, from which it follows that the circuit has a total area $A = O(n^4/r^2)$. Time satisfies the relation $T = O(r + \log(n/r))$, from which one has:

$$AT^2 = O(n^4), \text{ for } T \in [\Omega(\log n), O(n)]$$

As anticipated, this scheme enables one to reach the lower bound on AT^2 for any degree of parallelism.

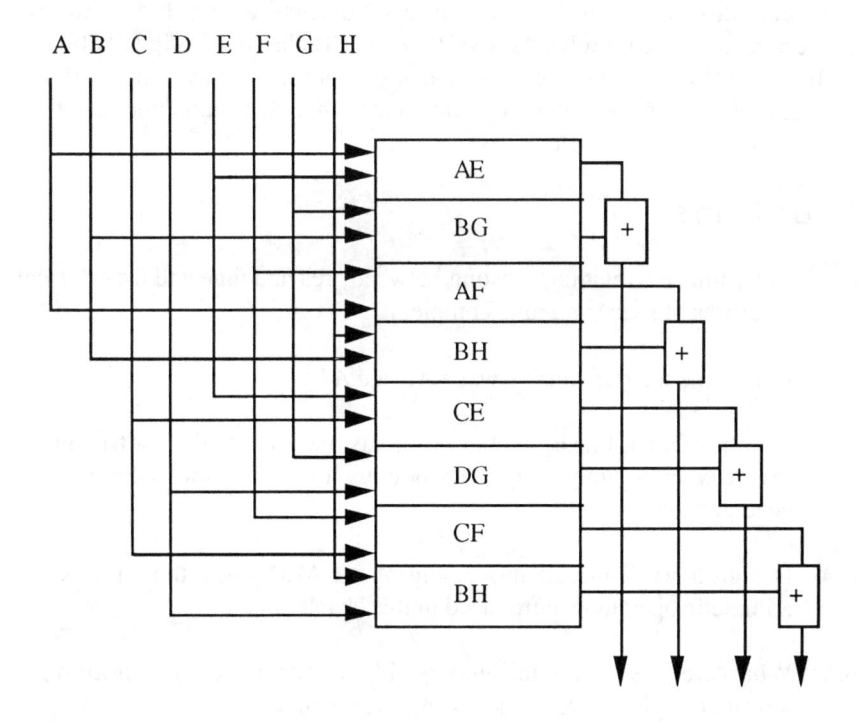

Figure 6.26 Circuit to execute the multiplication of p n/r x n/r matrices. Matrices are partitioned as 2 x 2 block matrices, with blocks (A, B, C, D) and (E, F, G, H).

Bibliographic notes

The fundamental notions on VLSI can be found in the textbook *Introduction to VLSI Systems* by Mead and Conway [MC80]. Concerning more specifically computational topics, the book by Ullman, *Computational Aspects of VLSI*, [Ul83] gives a complete introduction.

References to scientific papers published in the subject cannot but be incomplete. We will restrict ourselves to pointing out a few contributions.

Results on computational models are first of all contained in the PhD thesis by Thompson [Th80] and further in [AA80], [BPP81], [Ya81], [CM81]. Some questions related to the advantages one can have by replicating input data can be found in [KZ81].

Two rather complete reviews are [Ba83] and [Pr84].

The field of the VLSI complexity of specific computational problems is covered by a large number of papers. For arithmetic problems, one can see [BK81], [MP83], [MP85], [J80]. For combinatorial problems, one must point out the Ph.D thesis by Bilardi [Bi84] containing the study of the area–time complexity of sorting. This is also discussed in [JJ84]. Finally, concerning numerical problems, see [PV80], [CL87a], [CLR85], [CLR90].

In Chapter 6, we neglected, for reasons of brevity, many computational problems. The interested reader will find a more detailed panoramic review in the book by Ullman and references therein.

Problems

6.1 Determine the relations existing between area and time and the classical measures of space and time complexity.

6.2 Determine the relations between AT and AT^2.

6.3 State for which linear algebra problems one can obtain, for triangular matrices, area–time complexity bounds lower than those holding for general matrices.

6.4 Design a VLSI circuit implementing an ALU using the circuits for arithmetic operations introduced in this chapter.

6.5 What are the computational problems for which the notion of information flow gives non–trivial lower bounds?

6.6 Apply the techniques introduced in Section 6.2 to determine a lower bound for the computation of specific bilinear forms.

6.7 Determine lower complexity bounds for integer addition without using techniques based on flow or on the argument in Section 6.4.

6.8 Analyze the performance of a two–dimensional array when solving all numeric problems in Chapter 6.

6.9 Use the circuit for matrix multiplication to solve linear systems. (Hint. Apply Newton's method.) Which value of AT^2 can be obtained? Is it optimal?

Appendix A
Orders of Magnitude

Let f and g be two real-valued functions defined on the set of natural numbers:

$$f,g: \mathbf{N} \longrightarrow \Re$$

One says that:

1. f is of order *not greater* than g, denoting it by $f(n) = O(g(n))$, when $|f(n)| < \alpha |g(n)|$, where α is a constant not depending on n;
2. f is of order *not less* than g, denoting it by $f(n) = \Omega(g(n))$, if g is of order not greater than f, that is, if $g(n) = O(f(n))$;
3. f is of order *less* than g, denoting it by $f(n) = o(g(n))$, if:

$$\lim_{n \to \infty} \frac{f(n)}{g(n)} = 0$$

4. f is of the *same order* as g, denoting it by $f(n) = \theta(g(n))$, if:

$$\lim_{n \to \infty} \frac{f(n)}{g(n)} \in \Re$$

Appendix B
Number Representation

The notation used for numbers in scientific computations is the *floating-point* representation. It provides two pieces of information, called *mantissa* and *characteristic* (or *exponent* of the representation base), needed as a starting point to "reconstruct" the number itself. For instance, the notation $(11, -3)$ indicates the number $0.11*10^{-3}$ if the base of the representation system is 10, and if the position of the decimal point is assumed to be immediately to the left of the first non-zero digit of the mantissa. These last two pieces of information are implicit with respect to a specific representation system. For the sake of simplicity, we have adopted the decimal base; however, within a machine the base will always be 2 or a power of 2. Regarding the position of the decimal point, we adopt the universally accepted convention, that is, the decimal point is located immediately to the left of the first non-zero digit of the mantissa. In this way, possible significant zeros of a number (that is, zeros following a decimal point in the traditional positional representation) are not represented, whereas the relative information is contained in the exponent. A representation such as the above, where the first digit of the mantissa is different from 0, is called a *normalized* representation. For instance, the number $X = 0.00011$ can be represented in floating point by noting that:

$$0.00011 = 0.00011*10^0$$

or:

$$0.00011 = 0.11*10^{-3}$$

The first equality would give the representation $(00011,0)$, while the second one would lead to $(11, -1)$. As one can see, the second representation, being the normalized one, transfers to the exponent the information on the significant zeros. Within a computer the space allocated to the representation of a number is obviously finite. In general, it will be possible to represent t digits of the mantissa and s digits of the exponent. It is then apparent that a normalized notation allows one, for equal values of t and s, to represent numbers with greater precision. For instance, when $t = 4$, representing the number X starting from $0.00011*10^0$ causes the loss of a significant digit from the mantissa.

Let us now assume we must add two numbers represented in the (normalized) floating-point format, for instance $X = (m_1, e_1)$ and $Y = (m_2, e_2)$. The problem is to determine the normalized representation of the sum of two numbers. It is clear that the mantissa of the number $Z = X + Y$ is "in some way" the sum of the mantissas of X and Y. However, due to the normalization, the ith digit of the mantissa of X does not have, in general, the same importance as the ith digit of the mantissa of Y. It is then necessary to make the two mantissas comparable by means of an operation which is, in some sense, the inverse of normalization. If we agree on operating always on the number with the lower exponent, let it be X, this operation is performed as follows:

- the difference $d = e_2 - e_1$ is computed;
- zeros are inserted to the left of the mantissa of X;
- the exponent of X becomes e_2.

At this point the mantissa M of Z is obtained as the sum of the mantissas of X and Y and the exponent E of Z is e_2 (common to X and Y). Note however that M could not be normalized, and therefore a further normalization phase might be necessary, whereby leading zeros are deleted and the exponent is incremented. For instance, if X is taken as before and $Y = -0.27*10^{-4} = (-27, -4)$, one has $M = (+.11) + (-.027) = +.083$ and $E = -3$. Since M is not a normalized mantissa, it is necessary to adjust the result by deleting the leading zero and by decreasing E by one unit. The representation of the result will then be $(83, -4)$.

A discussion similar to the previous one also applies to multiplication (and to division). Therefore we can identify, in the execution of an arithmetic instruction, at least 4 phases, namely:

1. computation of the difference of the exponents;
2. mantissa line-up;
3. operation (sum or multiplication) on the mantissas;
4. result normalization.

Appendix C
Pidgin ALGOL Language

Pidgin ALGOL is not a standard programming language. It has been introduced in [AHU74], and subsequently it has been used, with various modifications, according to specific goals (see for instance [PS82]). The original language definition was not completely specified. In fact, there has never been the objective of implementing a Pidgin ALGOL compiler. The stated purpose was to have a flexible instrument to describe the fundamental ideas of an algorithm (although a rigourous one, allowing a precise analysis of the space and time complexity of an algorithm, without the need of equivalent RAM or Turing machine programs). Actually, Pidgin ALGOL allows *any kind of mathematical expression* which has a clear meaning and is amenable to a clear transformation into RAM (or other model) code. It further accepts all main types of data structures. In this book we used only a subset of the original Pidgin ALGOL commands. However, we have introduced new commands in order to describe aspects of parallel computations.

The *lexicon* of the language contains a few symbols having a predefined meaning. Such symbols will be outlined by a bold typeface. A Pidgin ALGOL program is then any command of the type indicated below (with a brief comment) and enclosed between a pair of program parentheses **begin end**.

1. VARIABLE ← EXPRESSION
   The effect of the command is to *assign* the value of EXPRESSION, computed in the context where the command appears, to VARIABLE. Note that it is the use of variables which makes context-dependent the value of an expression.
2. **if** CONDITION **then** COMMAND
   CONDITION must be any truth statement (that is, a statement with a value of *true* or *false*). If CONDITION is *true*, COMMAND is

executed followed by the command after 2, otherwise the command following 2 is directly executed.

3. **if** CONDITION **then** COMMAND-1 **else** COMMAND-2
CONDITION is as in command 2. If CONDITION is *true*, COMMAND-1 is executed, otherwise COMMAND-2. In both cases the command after 3 follows.

4. **while** CONDITION **do** COMMAND
CONDITION is as in command 2. If CONDITION is *true*, COMMAND is executed, then the program goes back (recursively) to command 4. If, at some point, CONDITION is false, the command following 4 is executed.

5. **for** VARIABLE ← VAL-INITIAL **until** VAL-FINAL **do** COMMAND
COMMAND is executed several times with VARIABLE taking all integer values between VAL-INITIAL and VAL-FINAL (included).

6. **begin**
COMMAND-1;
COMMAND-2;
. . .
COMMAND-*n*
end
The commands COMMAND-1, ..., COMMAND-*n* are executed in sequence.

7. **procedure** IDENTIFIER (PARAMETER-LIST); COMMAND
It is the procedural abstraction present (in various forms) in any programming language. Semantically it can be a command or an expression. In the latter case the last instruction executed by COMMAND must be **return** EXPRESSION and the value of EXPRESSION is the value of the procedure.

8. **return** EXPRESSION
If executed within a procedure it causes the end of the procedure itself and the transmission of EXPRESSION to the calling program.

9. **read** VARIABLE
A value is read from the storage medium and assigned to VARIABLE as a new value.

10. **write** EXPRESSION
The value of EXPRESSION is communicated to the outside world by means of the write medium.

11. **stop**
Halts program execution.

12. **comment**

Anything following the **comment** symbol, up to the first ";" character, is considered a comment and as such is ignored.

13. VARIABLE-L ← (VARIABLE-R **of** PORT)
    The value of the variable VARIABLE-R, of the processor specified by the communication port PORT, is transferred to the variable VARIABLE-L of the processor executing 13.

14. (VARIABLE-R **of** PORT) ← VARIABLE-L
    The value of the variable VARIABLE-L, of the processor executing 14, is transferred to the variable VARIABLE-R of the processor specified by the communication port PORT.

Bibliography

[AA80]   H. Abelson and P. Andreae, Information Transfer and Area–Time tradeoffs, *Comm. of the ACM*, **23** (1980), pp. 20-22.

[AHU74] A.V. Aho, J. Hopcroft and J.D. Ullman, *The Design and Analysis of Computer Algorithms*, Addison-Wesley, Reading, MA, 1974.

[AUY83] A.V. Aho, J.D. Ullman and M.Yannakakis, On Notions of Information Transfer in VLSI Circuits, Proc. of 15th Annual ACM Symp. on Theory of Computing, 1983, pp. 133-139.

[AKS83] M. Aitai, J. Komlos and E. Szemeredi, An O($n\log n$) Sorting Network, Proc. 15th Annual ACM Symposium on Theory of Computing, Boston, MA, 1983, pp. 133-139.

[At65]   A.J. Atrubin, An Interactive One-dimensional Real Time Multiplier, *IEEE Trans. on Computers*, **EC-14** (1965), pp. 394-399.

[AOES88] C. Aykanat, F. Ozguner, F. Ercal and P. Sadayappan, Iterative Algorithms for Solution of Large Sparse Systems of Linear Equations on Hypercubes, *IEEE Trans. on Computers*, **C-37** (1988), pp. 1554-1567.

[Ba78]   G.M. Baudet (1978), Asynchronous Iterative Methods for Multiprocessors, *J. Assoc. Comput. Mach.*, **25**(2), pp. 226-244.

[Ba81]   G.M. Baudet, On the Area Required by VLSI Circuits, in Kung, Sproull and Steele, eds., *VLSI Systems and Computations* (1981), pp. 100-107.

[Ba83]   G.M. Baudet, Design and Complexity of VLSI Algorithms, in *Foundations of Computer Science IV*, Mathematical Center, Amsterdam (1983), pp. 49-73.

[BCH84] P.W. Beame, S.A. Cook and H.J. Hoover, Log Depth Circuits for Division and Related Problems, Proc. 25th Annual IEEE Symposium on Foundations of Computer Science, Singer Island, FL, 1984, pp. 1-6. Also *SIAM J. Computing*, **15** (1986), pp. 994-1003.

Bibliography

[Be84] S.J. Berkowitz, On Computing the Determinant in Small Parallel Time Using a Small Number of Processors, *Inform. Process. Lett.*, **18** (1984), pp. 147-150.

[BBCM87] R. Bevilacqua, D. Bini, M. Capovani and O. Menchi, Introduzione alla Matematica Computazionale, Zanichelli, Bologna, G. Bilardi, The Area–Time Complexity of Sorting, PhD Thesis, University of Illinois at Urbana Champaign, 1984.

[Bi84] G. Bilardi, The Area–Time Complexity of Sorting, PhD Thesis, University of Illinois at Urbana Champaign, 1984.

[BPP81] G. Bilardi, M. Pracchi and F.P. Preparata, A Critique and an Appraisal of VLSI Models of Computation, in Kung, Sproull and Steele, eds., *VLSI Systems and Computations* (1981), pp. 81-88.

[BN86] G. Bilardi and A. Nicolau, Bitonic Sorting with $O(n\log n)$ Comparisons, 20th Annual Conf. on Inform. Sci. and Syst., Princeton, NJ, 1986.

[BP86] G. Bilardi and F.P. Preparata, Area–Time Lower-Bound Techniques with Applications to Sorting, *Algorithmica*, **1** (1986).

[BM75] A. Borodin and I. Munro, *The Computational Complexity of Algebraic and Numeric Problems*, American Elsevier, New York, 1975.

[Bo77] A. Borodin, On relating Time and Space to Size and Depth, *SIAM J. Computing*, **6** (1977), pp. 733-744.

[Bo82] A. Borodin, Structured vs. General Models in Computational Complexity, *Enseign. Math.*, **XXVIII** (1982), pp. 47-65.

[BGH82] A. Borodin, J. von zur Gathen and J. Hopcroft, Fast parallel Matrix and GCD Computations, *Information and Control*, **52** (1982), pp. 241-256.

[BH82] A. Borodin and J. Hopcroft, Routing, Merging and Sorting on Parallel Models of Computation, Proc. of 14th Annual ACM Symp. on Theory of Computing, San Francisco, CA, 1982, pp. 38-344. Also in *J. Comput. Sys. Sci.*, **30** (1985), pp. 130-145.

[BCP83] A. Borodin, S.A. Cook and N. Pippenger, Parallel Computation for Well-Endowed Rings and Space-Bounded Probabilistic Machines, *Information and Control*, **58** (1983), pp. 113-136.

[BG82] R.P. Brent and L.M. Goldschlager, Some Area–Time Tradeoffs for VLSI, *SIAM J. Computing*, **11** (1982), pp. 737-747.

[BK80] R.P. Brent and H.T. Kung, The Chip Complexity of Binary Arithmetic, Proc. of 12th Annual ACM Symp. on Theory of Computing, 1980, pp. 190-200.

[BK81] R.P. Brent and H.T. Kung, The Area–Time Complexity of Binary Multiplication, *J. Assoc. Comput. Mach.*, **28** (1981), pp. 521-534.

Bibliography 259

[BK82] R.P. Brent and H.T. Kung, A Regular Layout for Parallel Adders, *IEEE Trans. on Computers*, **C-31** (1982), pp. 260-264.

[CM69] D. Chazan and W. Miranker (1969), Chaotic Relaxation, *Linear Algebra Appl.*, **2**, pp. 199-222.

[CM81] B. Chazelle and L. Monier, A Model of Computation for VLSI with Related Complexity Results, Proc. of 13th Annual ACM Symp. on Theory of Computing, Milwaukee, 1981, pp. 318-325.

[CL87a] B. Codenotti and G. Lotti, A VLSI Fast Solver for Tridiagonal Linear Systems, *Inform. Process. Lett.*, (1987), pp. 111-114.

[CL87b] B. Codenotti and G. Lotti, Area–Time Tradeoffs for Bilinear Forms Computations in VLSI, *Inform. Process. Lett.*, (1987), pp. 107-109.

[CLR85] B. Codenotti, G. Lotti and F. Romani, VLSI Implementation of Iterative Methods for the Solution of Linear Systems, Integration, *The VLSI Journal*, **3** (1985), pp. 211-221.

[CLR90] B. Codenotti, G. Lotti and F. Romani, Area–Time Complexity of Matrix–Vector Multiplication, *Journal of Parallel and Distributed Computing*, (1990), in press.

[Co71] S.A. Cook, The Complexity of Theorem Proving Procedures, Proc. of 3rd Annual ACM Symp. on Theory of Computing, 1971, pp. 151-158.

[Co74] S.A. Cook, An Observation on Time–Storage Trade-off, *J. Comput. Sys. Sci.*, **9** (1974), pp. 308-316.

[Co81] S.A. Cook, Towards a Complexity Theory of Synchronous Parallel Computation, *Enseign. Math.*, **XXVII** (1981), pp. 99-124.

[Co85] S.A. Cook, A Taxonomy of Problems with Fast Parallel Algorithms, *Information and Control*, **64** (1985), pp. 2-22.

[Co86] R. Cole, Parallel Merge Sort, Proc. 27th Annual IEEE Symposium on Foundations of Computer Science, 1986, pp. 511-516.

[Cs76] L. Csanky, Fast Parallel Matrix Inversion Algorithms, *SIAM J. Computing*, **5** (1976), pp. 618-623.

[Da65] L. Dadda, Some Schemes for Parallel Multipliers, *Alta Frequenza*, **34** (1965), pp. 343-356.

[DK89] E. Dahlaus and M. Karpinski, An Efficient Parallel Algorithm for the Minimal Elimination Ordering (MEO) of an Arbitrary Graph, Proc. 30th Annual IEEE Symposium on Foundations of Computer Science, Raleigh, NC, 1989, pp. 454-461.

[DER86] I.S. Duff, A.M. Erisman and J.K. Reid, *Direct Methods for Sparse Matrices*, Oxford Science Publications, 1986.

[EG88] D. Eppstein and Z. Galil, Parallel Algorithmic Techniques for Combinatorial Computation, *Annual Review of Computer Science*, **3** (1988), pp. 233-283.

[FF63]   D.K. Faddeev and V.N. Faddeeva, *Computational Methods of Linear Algebra*, W.H. Freeman and Company, San Francisco, 1963.

[FIS88] D.C. Fisher, Your Favorite Parallel Algorithms Might Not Be as Fast as You Think, *IEEE Trans. on Computers*, **C-37** (1988), pp. 211-213.

[FLY66] M.J. Flynn, Very High-Speed Computing Systems, *Proc. of the IEEE*, **54** (1966), pp. 1901-1909.

[FLY72] M.J. Flynn, Some Computer Organizations and Their Effectiveness, *IEEE Trans. on Computers*, **C-21** (1972), pp. 948-960.

[FW78]   S. Fortune and J. Wyllie, Parallelism in Random Access Machines, Proc. of 10th Annual ACM Symp. on Theory of Computing, 1978, pp. 114-118.

[GP81]   Z. Galil and W.J. Paul, An Efficient General Purpose Parallel Computer, Proc. of 13th Annual ACM Symp. on Theory of Computing, Milwaukee, 1981, pp. 247-262.

[GJ79]   M.R. Garey and D.S. Johnson, *Computers and Intractability: A Guide to the Theory of NP-Completeness*, Freeman and Company, San Francisco, CA, 1979.

[Ga86]   12th Int. Symp. on Math. Foundations of Comput. Science, *Lecture Notes in Computer Science*, **233**, Springer-Verlag, New York, 1986, pp. 93-122.

[GEN78] W.M. Gentleman, Some Complexity Results for Matrix Computations, *J. Assoc. Comput. Mach.*, **25** (1978), pp. 112-115.

[GIL87] W.K. Giloi, Interconnection Networks for Massively Parallel Computer Systems, in Treleaven, P., and Vanneschi, M. (cur.), Future Parallel Computers, *Lecture Notes in Computer Science*, **272**, Springer-Verlag, Berlin, 1987.

[GL78]   L.M. Goldschlager, A Unified Approach to Models of Synchronous Parallel Machines, Proc. of 10th Annual ACM Symp. on Theory of Computing, 1978, pp. 89-94.

[GL82]   L.M. Goldschlager, A Universal Interconnection Pattern for Parallel Computers, *J. Assoc. Comput. Mach.*, **29** (1982), pp. 1073-1086.

[GL83]   L.M. Goldschlager and A.M. Lister, *Computer Science: A Modern Introduction*, Prentice-Hall, Englewood-Cliffs, NJ, 1983.

[Gu90]   I. Guessarian (ed.), Semantics of Systems of Concurrent Processes, *Lecture Notes in Computer Science*, **469**, Springer-Verlag, Berlin, 1990.

[GVL83] G.H. Golub and C.F. Van Loan, *Matrix Computations*, J. Hopkins University Press, Baltimore, MD, 1983.

Bibliography

[HH81] R. Haggkvist and P. Hell, Parallel Sorting with Constant Time for Comparisons, *SIAM J. Computing*, **10** (1981), pp. 465-472.

[HLS65] J. Hartmanis, P.M. Lewis II and R.E. Stearns, Hierarchies of Memory Limited Computations, Proc. 6th Annual IEEE Symp. on Switching Circuit Theory and Logical Design, pp. 179-190.

[He78] D. Heller, A Survey of Parallelism in Numerical Linear Algebra, *SIAM Rev.*, **20** (1978), pp. 740-777.

[HST76] D. Heller, D.K. Stevenson and J.F. Traub, Accelerated Iterative Methods for the Solution of Tridiagonal Systems on Parallel Computers, *J. Assoc. Comput. Mach.*, **23** (1976), pp. 636-654.

[HCS79] D.S. Hirschberg, A.K. Chandra and D.V. Sarwate, Computing Connected Components on Parallel Computers, *Comm. of the ACM*, **22** (1979), pp. 461-464.

[Ho64] A.S. Householder, *The Theory of Matrices in Numerical Analysis*, Blaisdell Pu. Co., New York, 1964.

[HU79] J.E. Hopcroft and J.D. Ullman, *Introduction to Automata Theory, Languages and Computations*, Addison-Wesley, Reading, MA, 1979.

[HWB84] K. Hwang and F. Briggs, *Computer Architecture and Parallel Processing*, McGraw-Hill, New York, 1984.

[JJ84] J. Ja'Ja', The VLSI Complexity of Selected Graph Problems, *J. Assoc. Comput. Mach.*, **31** (1984).

[JL77] N.D. Jones and W.T. Laaser, Complete Problems for Deterministic Polynomial Time, *Theoretical Computer Science*, **3** (1977), pp. 105-117.

[Jo75] N.D. Jones, Space-bounded Reducibility among Combinatorial Problems, *J. Comput. Sys. Sci.*, **11** (1975), pp. 68-85.

[J80] R.B. Johnson, The Complexity of a VLSI Adder, *Inform. Process. Lett.*, **11** (1980), pp. 92-93.

[KO62] A. Karazuba and Y. Ofman, Multiplication of Multidigit Numbers on Automata, *Dokl. Akad. Nauk SSSR*, **145** (1962), pp. 293-294.

[Ka72] R.M. Karp, Reducibility among Combinatorial Problems, in R.E. Miller and J.W. Thatcher (eds.), *Complexity of Computer Computations*, Plenum Press, New York, 1972.

[KR90] R.M. Karp and V. Ramachandran, Parallel Algorithms for Shared-Memory Machines, in *Handbook of Theoretical Computer Science*, North Holland, Amsterdam, 1990.

[KZ81] Z.M. Kedem and A. Zorat, On Relations between Input and Communication/Computation in VLSI, Proc. 22nd Annual IEEE Symposium on Foundations of Computer Science, 1981, pp. 37-44.

[Ku76] H.T. Kung, New Algorithms and Lower Bounds for the Parallel Evaluation of Certain Rational Expressions and Recurrences, *J. Assoc. Comput. Mach.*, **23** (1976), pp. 252-261.

[Ku82]　H.T. Kung, Why Systolic Architectures?, *Computer Magazine*, **15** (1982), pp.37-46.

[KL80]　H.T. Kung and C.E. Leiserson, Systolic Arrays (for VLSI), in C.A. Mead and L. Conway, *An Introduction to VLSI Systems*, Addison-Wesley, Reading, MA,1980.

[LF80]　R.E. Ladner and M.J. Fischer, Parallel Prefix Computation, *J. Assoc. Comput. Mach.*, **27** (1980), pp. 831-838.

[Le81]　F.T. Leighton, New Lower Bound Techniques for VLSI, Proc. 22nd Annual IEEE Symposium on Foundations of Computer Science, 1981, pp. 1-12.

[Le91]　F.T. Leighton, *Introduction to Parallel Algorithms and Architectures: Arrays, Trees and Hypercubes*, Morgan Kaufmann, 1991.

[Lu82]　F. Luccio, La struttura degli algoritmi, Boringhieri, Turin, 1982.

[LV83]　W.E. Luk and J. Vuillemin, Recursive Implementation of Optimal Time VLSI Integer Multipliers, VLSI 83, Trondheim, Norway, 1983.

[MC80]　C.A. Mead and L. Conway, *Introduction to VLSI Systems*, Addison-Wesley, Reading, MA, 1980.

[Me84]　K. Mehlhorn, Optimal AT^2 Integer Division and Square Rooting, Integration, *The VLSI Journal*, **2** (1984), pp. 163-167.

[MP83]　K. Mehlhorn and F.P. Preparata, Area–Time Optimal VLSI Integer Multiplier with Minimum Computation Time, *Information and Control*, **58** (1983), pp. 137-156.

[MP85]　K. Mehlhorn and F.P. Preparata, Area–Time Optimal Division, Tech. Rep. ACT-63, Coordinated Science Laboratory, University of Illinois at Urbana-Champaign. Agust 1985.

[Me78]　J.D. Meindl, Gli Elementi dei Circuiti Microelettronici, *Le Scienze*, **123** (1978), pp. 16-27.

[MR85]　G.L. Miller and J. Reif, Parallel Tree Contraction and its Applications, Proc. 26th Annual IEEE Symp. on Foundations of Computer Science, 1985, pp. 478-489.

[Mu63]　D.E. Muller, Asynchronous Logic and Application to Information Processing, in Aiken and Main (Eds.), *Switching Theory in Space Technology*, Stanford University Press, Stanford, CA, 1963.

[MP75]　D.E. Muller and F.P. Preparata, Bounds to Complexities of Networks for Sorting and Switching, *J. Assoc. Comput. Mach.*, **22** (1975), pp. 195-201.

[Mu86]　K. Mulmuley , A Fast Parallel Algorithm To Compute the Rank of a Matrix over an Arbitrary Field, Proc. of 18th Annual ACM Symp. on Theory of Computing, Berkeley CA, 1986, pp. 338-339.

[NS81] D. Nassimi and S. Sahni, Data Broadcasting in SIMD Computers, *IEEE Trans. on Computers*, **C-30** (1981), pp. 101-107.

[PAW86] D.A. Padua and M.J. Wolfe, Advanced Compiler Optimization for Supercomputers, *Comm. of the ACM*, **29** (1986), pp. 1184-1201.

[PR85] V. Pan and J. Reif, Efficient Parallel Solution of Linear Systems, Proc. of 17th Annual ACM Symp. on Theory of Computing, 1985, pp. 143-152.

[PS82] C.H. Papadimitriou and K. Steiglitz, *Combinatorial Optimization*, Prentice-Hall, Englewood Cliffs, NJ, 1982.

[Pa87] J. Parberry, *Parallel Complexity Theory*, Pitman, London, 1987.

[Pe67] M.C. Pease, Matrix Inversion Using Parallel Processing, *J. Assoc. Comput. Mach.*, **14** (1967), pp. 757-764.

[Pi79] N. Pippenger, On Simultaneous Resource Bounds, Proc. 20th Annual IEEE Symposium on Foundations of Computer Science, 1979, pp. 307-311.

[PF79] N. Pippenger and M.J. Fischer, Relations Among Complexity Measures, *J. Assoc. Comput. Mach.*, **26** (1979), pp. 361-381.

[Pr84] F.P. Preparata, VLSI Algorithms and Architectures, Proc. of 11th Symp. on the Mathematical Foundations of Computer Science, 1984, pp. 149-161.

[PS78] F.P. Preparata and D.V. Sarwate, An Improved Parallel Processor Bound in Fast Matrix Inversion, *Inform. Process. Lett.*, **7** (1978), pp. 148-149.

[PV80] F.P. Preparata and J. Vuillemin, Area–Time Optimal VLSI Networks for Multiplying Matrices, *Inform. Process. Lett.*, **11** (1980), pp. 77-80.

[PV81] F.P. Preparata and J. Vuillemin, The Cube-Connected-Cycles: A Versatile Network for Parallel Computation, *Comm. of the ACM*, **24** (1981), pp. 300-309.

[Qu87] M.J.Quinn, *Designing Efficient Algorithms for Parallel Computers*, McGraw-Hill (1987).

[Re86] J.H. Reif, Logarithmic Depth Circuits for Algebraic Functions, *SIAM J. Computing*, **15** (1986), pp. 231-242.

[Re81] R. Reischuk, A Fast Probabilistic Parallel Sorting Algorithm, Proc. 22nd Annual IEEE Symposium on Foundations of Computer Science, 1981, pp. 212-219.

[RET86] R. Rettberg and R. Thomas, Contention is no Obstacle to Shared-Memory Multiprocessing, *Comm. of the ACM*, **29** (1986), pp. 1202-1212.

[Ru81] W.L. Ruzzo, On Uniform Circuit Complexity, *J. Comput. Sys. Sci.*, **22** (1981), pp. 365-383.

[SB77] A.H. Sameh and R.P. Brent, Solving Triangular Systems on a Parallel Computer, *SIAM J. Numerical Analysis*, **14** (1977), pp. 1101-1113.

[SK78] A.H. Sameh and D.J. Kuck, On Stable Parallel Linear System Solvers, *J. Assoc. Comput. Mach.*, **25** (1978), pp. 81-91.

[Sa76] J.E. Savage, *The Complexity of Computing*, John Wiley, New York, 1976.

[Sa79] J.E. Savage, Area–Time Tradeoffs for Matrix Multiplication and Transitive Closure in the VLSI Model, Proc 17th Allerton Conference, 1979, pp. 670-676.

[Sa81] J.E. Savage, Area–Time Tradeoffs for Matrix Multiplication and Related Problems in VLSI Models, *J. Comput. Sys. Sci.*, **20** (1981), pp. 230-242.

[Sa70] W. Savitch, Relationships Between Nondeterministic and Deterministic Tape Complexities, *J. Comput. Sys. Sci.*, **4** (1970), pp. 177-192.

[SV81] Y. Shiloach and U. Vishkin, Finding the Maximum, Merging and Sorting in a Parallel Computation Model, *Journal of Algorithms*, **2** (1981), pp. 88-102.

[Si79] H.J. Siegel, A Model of SIMD Machines and a Comparison of Various Interconnection Networks, *IEEE Trans. on Computers*, **C-28** (1979), pp. 907-917.

[SV84] L. Stockmeyer and U. Vishkin, Simulation of Parallel Random Access Machines by Circuits, *SIAM J. Computing*, **13** (1984), pp. 409-422.

[STO71] H.S. Stone, Parallel Processing with the Perfect Shuffle, *IEEE Trans. on Computers*, **C-20** (1971), pp. 153-161.

[STO75] H.S. Stone (ed.), *Introduction to Computer Architecture*, Science Research Associates INC, Chicago, 1975.

[St69] V. Strassen, Gaussian Elimination is not Optimal, *Numerische Matematik*, **13** (1969), pp. 354-356.

[SM77] I.E. Sutherland and C.A. Mead, Microelectronics and Computer Science, *Scientific American*, Sept. 1977, pp. 210-228.

[Th79] C.D. Thompson, Area–Time Complexity for VLSI, Proc. of 11th Annual ACM Symp. on Theory of Computing, 1979, pp. 81-88.

[Th80] C.D. Thompson, *A Complexity Theory for VLSI*, PhD Thesis, Carnegie Mellon University, Pittsburg, 1980.

[Tr73] J.F. Traub, Iterative Solution of Tridiagonal Systems on Vector and Parallel Computers, in J.F. Traub (ed.), *Complexity of Sequential and Parallel Numerical Algorithms*, Academic Press, New York, 1973.

[TV86] P. Treleaven and M. Vanneschi (eds.), Future Parallel Computers, *Lecture Notes in Computer Science*, **272**, Springer-Verlag, Berlin, 1986.

[Ul83]   J.D. Ullman, *Computational Aspects of VLSI*, Computer Science Press, 1983.
[Va75]   L.G. Valiant, Parallelism in Comparison Problems, *SIAM J. Computing*, **4** (1975), pp. 348-355.
[VB81]   L.G. Valiant and G.J. Brebner, Universal Schemes for Parallel Communication, Proc. of 13th Annual ACM Symp. on Theory of Computing, 1981, pp. 263-277.
[Vu83a] J. Vuillemin, A Combinatorial Limit to the Computing Power of VLSI Circuits, *IEEE Trans. on Computers*, **C-32** (1983), pp. 147-150.
[Vu83b] J. Vuillemin, A Very Fast Multiplication Algorithm for VLSI Implementation, Integration, *The VLSI Journal*, **1** (1983), pp. 33-52.
[VW83]   U. Vishkin and A. Wigderson, Trade-offs Between Depth and Width in Parallel Computation, Proc. 24th Annual IEEE Symp. on Foundations of Computer Science, 1983, pp. 146-153.
[Wa64]   C.S. Wallace, A Suggestion for a Fast Multiplier, *IEEE Trans. on Computers*, **EC-13** (1964), pp. 14-17.
[WF84]   C. Wu and T.Feng, *Interconnection Networks for Parallel and Distributed Computations*, IEEE Computer Society, 1984.
[Ya81]   A.C. Yao, The Entropic Limitations on VLSI Computations, Proc. of 13th Annual ACM Symp. on Theory of Computing, Milwaukee, 1981, pp. 308-311.
[Zi86]   M. Venturini Zilli (ed.), Mathematical Models for the Semantics of Parallelism, *Lecture Notes in Computer Science*, **280**, Springer-Verlag, Berlin, 1986.

Index

A
A, see Register, storage
Acceptance (status), 47
Addition
   integer, 73, 75, 77, 85, 131, 231-234
   iterated, 85
Algorithm
   Csanky's, 152-155
   non-deterministic, 10-11
   optimal, 26
   parallel, 69, 124-165
   probabilistic, 10-11
Allocation, processor, 115
Alphabet, 8
   in a Turing Machine, 20
   input, 20
Antidependence, 172
Architecture, parallel
   common memory, 199-206
   distributed memory, 116, 204-206
   message-passing based, see Architecture, parallel, distributed memory
Area, of a circuit, 208, 214, 217
Area-Time, see, Complexity, area-time
Array
   linear, 189, 240, 244
   processor, 175-180
   square, see Mesh
   systolic, 240-241
AT, see Complexity, area-time
AT2, see Complexity, area-time

B
Base, of a circuit, 70
Behaviour, asymptotic, of an algorithm, 40-41
Bisection, 224
Boundary chip, 215
Bound, lower complexity, 42-43, 80, 103, 110, 115, 216-231
Bound, upper complexity, 42
Bus, 189
Butterfly network, 193

C
Cayley-Hamilton (theorem), 153
Chip
   (boundary), 215
   (general), 215
Circuit
   arithmetic, 107
   boolean, 68-76
   integrated, see Circuit, VLSI
   synchronous, 76
Circuit, VLSI, 208-216
Class, complexity, 48-50, 54
Classification, according to Flynn, 170
Clique, 29
CLIQUE, see Problem, maximal determination
CLIQUE(K), see Problem, k-clique determination
Closure, of a graph, 52-53, 85
CMOS, 211
Communication among processes, see Interaction, direct
Complexity
   area-time, 216-231
   computational, 24-26
      concrete, 26, 33-47
   of an interconnection structure, 188
   of a problem, 41-43
   of a RAM, 44-45

of a Turing Machine, 44-45
Component, connected, of a graph, 142
Computability, see theory of computability
Computation, 6, 9, 34-36
  convergent, 7
  deterministic, 9
  divergent, 7
  non-deterministic, 10
  of a network, 118
  of a PRAM, 88
  of a RAM, 14-16
  of a Turing Machine, 20-22
  parallel, 63, 67
  probabilistic, 10
Computation, prefix, 137-138
Computation, suffix, 139-140
Computer, conventional, 6-7, 9
Conditions, Bernstein's, 65
Configuration, of a Turing Machine, 21, 26
Cost
  average case, 37-40
  bit, 69
  logarithmic, 36
  of a computation, 34-36
  of an algorithm, 34-40
  of a VLSI circuit, 216
  unitary, 34-37
  worst case, 37-40
Counter, Dadda's, 133
CPU, see Unit, central processing
CRCW, see Machine, random access parallel CRCW
CREW, see Machine, random access parallel CREW
Crossbar, 193-195
Cube network, 193-194
Cube, boolean, see Hypercube

D
DCLIQUE(K) see Problem, existence of a k-clique
Deadlock, 184
Degree
  of integration, 209
  of interconnection, 118, 120-121
  of a rational function, 110
Dependence
  cyclic, 184
  flow, 172
  functional, 227
  output, 172
Depth
  of a boolean circuit, 70, 76
  of a family of arithmetic circuits, 108
  of a family of boolean circuits, 74, 80
  of a parallel comparison tree, 113
Description, instantaneous, 21
Determinant, 32, 113, 158, 161
Device, input/output, 6-7, 9
Device, I/O see Device, input/output
Digraph, see Graph, direct
Diode, 210
Distance, between processors, 120
Divide et impera, 124-125
Division, integer, 85, 86
Doping, 209
Doubling, recursive, 127

E
Efficiency, 67, 129-130
  polynomial (criterion), 46
Element, processing, 189
EREW, see Machine, random access EREW
Evaluation, expression, 139
Event, 181-185

Index

F
F, see Register, status indicator
Factorization
   LDU, 126
   QR, 150
Family of circuits
   arithmetic, 108
      uniform, 111
   boolean, 73-78
      uniform, 77-83
Fan-in, 70
Fan-out, 70
Fill-in, 143
Flow
   data, 170-171
   information (in a VLSI circuit), 218
   instruction, 171-172
   internal, 218
   I/O, 218
   primary, see Flow, I/O
   secondary, see Flow, I/O
Forest, 29
FORK, 186
Form, bilinear, 229
Function, 9
   computable, 25-26
   computed
      by an algorithm, 9
      by a boolean circuit, 72
      by a family of arithmetic, circuits, 108
      by a PRAM, 88
      by a RAM, 16, 24
      by a Turing Machine, 22, 25
   interconnection, 118
   non-computable, 25
   partial, 9
   program interpretation, 25
   recursive, see Function, computable
   space-constructible, 48
   total, 9, 26

transition, 20
transitive, 229

G
GAP, see Problem, graph accessibility
Graph, 28-32, 50
   acyclic, 28
   connected, 29
   direct, 28-31
   interconnection, 119
   labelled, 28
   of the computation, 52, 126-128
   oriented, see Graph, direct
   partial ordering, 62-63
Grid (model), 215, 222
   primary, 222
   secondary, 222
Group of permutations, 229
Group, transitive, 229

H
HAMILTON, see problem, Hamilton paths
Hardware (computation resource), 60
Hierarchy, space (theorem), 49
Hypercube, 193-195, 196-199

I
Identity (Newton's), 152
Indicator, status, see Event
Instance, of a problem, 27, 29-33, 37
Instruction, vector, 172
Interaction
   direct process, 180-186
   in common memory architectures, 180-184
   in distributed memory architectures, 184
   indirect, 180
Interconnection, complete, 193

Inversion, matrix, 149, 152-158
Iterations, vector, 126

J
JOIN, 187
Junction, *p-n*, 210

K
k-clique, see Clique

L
Language, 8-9
  accepted by a Turing Machine, 47
  context-free, 60, 85
  formal, 9-11
  programming, 8, 60, 65
  recognized by a boolean circuit, 74
  recognized by an algorithm, 32
Layout, 208, 214
Length of a computation, 34
List, 138

M
Machine, random access, 11-19, 35-36, 44-47
  parallel, 86-90, 94-101, 116
    CRCW, 99, 102-105, 117
    CREW, 99, 102-104, 116
    EREW, 99, 101-104, 106, 116
Machine, Turing, 19-25, 44-49, 52
  non-deterministic, 24, 50-52
  universal, 25
  with oracle, 53
Machine, Von Neumann, 2, 7
Matrix, 16
  adjacency, 28
  adjoint, 112, 158, 161
  sparse, 142-144
  symmetric, 28
  triangular, 153-155
Memory, 67, 168-169, 200-202
Merging, 144-145
Mesh, 189, 240
  of trees, 243-244
Method
  direct, 126
  elimination, 149-152
  Gauss-Jordan, 149
  Givens, 149
  iterative, 126, 136, 161-164, 196-199
    asynchronous, 162-164
  Newton, 135-136, 161-162
Microelectronics, 208
MIMD, 171, 179, 188
Minimum degree, 143
Model, computation, 8, 11-26, 44-47
  algorithmic, 66
  general, 106
  parallel, 59-123
  structured, 106-107
  VLSI, 213-216
    synchronous, 215
Module, basic, 243-244
MOSFET, 210
Multiplication
  integer, 14, 85, 131-135, 235-238
  iterated, 85, 135-137
  matrix, 16, 37, 85, 112, 244-249
  polynomial, 112
  row by column, see Multiplication, matrix
Multiprocessor
  common memory, 199
  distributed memory, 199

N
NC, 84, 106
  arithmetic, 112
NC^k, 84, 106
  arithmetic, 112
Network, see Network, fixed
    interconnection structure
  fixed interconnection
    structure, 116-121
  uniform, 120
  one stage, 189-190
  permutation, 178
NMOS, 210
Non-determinism, 10, 16, 26, 50
Non-redundancy (principle), 71
NP, 50, 52-54
NP-complete (problem), 26, 53-55, 143
NPSPACE, 52
Number of parallel steps
  (computation resource), 66
Number of processors
  (computation resource),
    66-68, 114, 118

O
Ordering, see Problem,
  ordering
Organization, memory
  (interlaced), 202
  (sequential), 202

P
P, 50, 53-56
P=? NP, 26, 53-54
Parallelism, limited, 67
Paths, 28
  Hamilton, 29
Path System Accessibility, 56
PC, see Register, program
  counter
P-complete (problem), 55, 60
Perfect shuffle, 189

Pidgin Algol, 9, 254-256
Pipeline, 169-170
Place-determinate, see
  Protocol, place-determinate
Pointer jumping, 139
POLYLOGSPACE, 50
Polynomial, characteristic,
  112, 152-154, 158-161
Port, communication, 117, 188
Port, I/O, 214
PRAM, see Machine, parallel
  random access
Problem, 27, 29-33, 216
  computational, see
    Problem, decision
  decision, 32, 47, 54
  existence of a k-clique,
    29-30
  graph accessibility, 33, 50, 85, 100
  Hamilton paths, 29-31, 55
  k-clique determination,
    29-31
  language recognition,
    32-33, 47, 85
  maximal k-clique
    determination, 55
  maximum determination,
    114-115
  optimization, 55
  ordering, 18, 31, 104,
    114-116, 144-149
  program termination, 25
  sparse, 142
Process, 3, 179, 184
  executing, 184
  ready for execution, 184
  (status transition), 184
  waiting
Processor, 64, 168-169
  hexagonal, 240
  square, 240
  vector, 170, 172, 179
  virtual, see Process

Program, 8
  straight-line, 18
Programming, synchronous
  parallel, 2
Protocol
  I/O, 215
  place-determinate, 215
  semellective, 215
  time-determinate, 215
  unilocal, 215
  word-local, 215
Pseudoadder, 131-132
PSPACE, 50, 52, 95

Q
Queue, waiting, 184

R
RAM, see Machine, random
  access
Rank, of a matrix, 161
RECEIVE, 185
Reducibility
  in logarithmic space, 55
  in polynomial time, 54
  NC^1, 85
Reduction of the number of
  tapes (theorem), 47-48
Reduction of the number of
  processors, 128-130
Register, 6
  accumulator, 12, 87
  program counter, 11, 86
  status indicator, 87
Reject (state), 47
Relationship, 27
  polynomial, 46
    complexity, 46
Resource, computing, 11, 25,
  34, 59-60, 66
Resources, simultaneous, 83,
  94, 96, 104
Ring, 189

S
Satisfiability, 54
Saturation, 225-227
Savitch (theorem), 51-52
Scaling, capability, 188
Semantics, 9
  functional, 9
  operational, 9
Semellective, see Protocol,
  semellective
Semiconductor, 208-211
SEND, 185
Separator, 221-224
Sequence
  bitonic, 146
  unimodal, 146
Set, instruction, 6
  of a network, 117
  of a random access
    machine, 12-14, 46-47
  of a random access parallel
    machine, 87, 97-99
Shuffle-exchange, 208
Shunt, 140
SIGNAL, 181-182
Silicon, 208-209
SIMD, 2, 171-179
Size
  of a boolean circuit, 70, 76
  of a family of arithmetic
    circuits, 108
  of a family of boolean
    circuits, 74, 83
  of a problem, 29
Sorting, see Problem, bitonic
  ordering
Space (computation resource),
  7, 34-36, 81, 93
Speed-up, 67
State, 6, 20, 34
String, 8
Structure, interconnection,
  118, 177, 187-199
Sum, prefix, 104, 137
Supercomputer, 169-170, 174

Synchronization, 179, 183
Syntax, 8
System, linear, 142, 149-150, 152

T
Technology, VLSI, 207-213, 239-240
Tensor, 230
Termination, program (problem), 25
Tessellation, 222-223, 225
Theory, computability, 24-26
Thesis
  of Church-Turing, 25-26
  of parallel computation, 60, 83, 93
  of sequential computation, 46-47
Time (computation resource), 7, 34-36, 59, 66, 83, 93, 216-218
  communication, 7
  computing, 7
  transmission (in a VLSI circuit), 215
Time-determinate, see Protocol, time-determinate

Transistor, 209-212
Tree, 28
  binary, 29, 43, 189
  comparison, see decision tree
  decision, 18
  elimination, 143
  parallel, 113
  Wallace, 131

U
Uniformity, 78, 93, 110, 119
Unilocal, see Protocol, unilocal
Unit, central processing, 6-7

V
Variable, I/O, 219-221, 224, 226-227
Vectorization, 126, 172-174

W
WAIT, 181-182
Width
  of a circuit, 76
  of a family of circuits, 83
Word-local , see Protocol, word-local